Currents in Submarine Canyons and Other Seavalleys

AAPG Studies in Geology No. 8

Currents in Submarine Canyons and Other Seavalleys

By

Francis P. Shepard, Neil F. Marshall, Patrick A. McLoughlin, and Gary G. Sullivan

Geological Research Division
Scripps Institution of Oceanography

Published By
The American Association of Petroleum Geologists
Tulsa, Oklahoma, U.S.A., 1979

Published August 1979
Library of Congress Catalog Card No. 79-53686
ISBN: 0-89181-012-9

The AAPG staff responsible:

John Shelton, editor
Gary Howell, editorial
Ronald Hart, editorial
Carol Short, production

Printed by Edwards Brothers, Inc.
Ann Arbor, Michigan, U.S.A.

The American Association of Petroleum Geologists

Table of Contents

Acknowledgements

Throughout the 10 years of this study, support has been provided principally by National Science Foundation Grants GA 826, GA 11463, GA 19492, OCE74-22089, and OCE77-17570, and by Office of Naval Research Contracts USN N00014-69-A-0200-6006 and USN N00014-75-C-0152.

We are grateful to the many geologists who arranged for our use of the ships and boats of their institutions and agencies so that we could set our current meters in many environments in various parts of the world. These include Donn Gorsline, George Keller, K. O. Emery, Robert Dill, Erk Reimnitz, David Drake, Gary Greene, P. J. Fischer, Orrin Pilkey, Robert Andrews, John Milliman, J. L. Luternauer, James Trumbull, and John Warme.

Many suggestions were made by Charles Nordstrom, George Keller, and Paul Komar, who critically read the manuscript. Good suggestions also were provided by K. O. Emery and R. E. Flick. We appreciate the help of Nance North who combed the literature for the bibliography and carefully copied the manuscript. The assistance of Margaret Miller during the early years of this study is greatly appreciated. Thanks are also due to those who worked on the project in our offices: John Roselius, Mary Moore, and Tom Harper. We thank Elizabeth Shepard for listening to readings of the manuscript and offering helpful suggestions. For physical oceanographic aspects of the work, we are grateful for suggestions from Charles Cox, and also from Linda Holmes who has been preparing a thesis using our data and will publish those results separately.

We are especially grateful to the officers and crews of many ships for their able assistance during deployment and recovery operations, especially Woody Reynolds, Garrett Coleman, and Albert Arsenault. The pilots, Larry Shoemaker and Don Saynor, and the support crew of submersible *Deep Quest* also were very helpful, as were Richard Slater and James Vernon of the submersible *Nekton Gamma* and the officers and crew of the submersible *Seacliff*.

We also thank those many friends, technicians, secretaries, and crew members who assisted us during our numerous operations, especially Ted Gustafson, who helped with local deployments of the equipment.

Currents in Submarine Canyons and Other Seavalleys[1]

Francis P. Shepard, Neil F. Marshall, Patrick A. McLoughlin, and Gary G. Sullivan[2]

Abstract——It is a widely held opinion that submarine canyons were cut during the glacial stages of low sea level and now are essentially dormant features, disturbed only on rare occasions by high-speed turbidity currents of torrential proportions. Our research contradicts this concept of the canyons, citing nine years of measurements of currents in the axes of canyons and in other types of seavalleys. Approximately 200 records from continuously operating current meters (during periods of several days to as much as a month) have shown that currents rarely cease flowing alternately up and down the floor of the valleys and frequently attain speeds sufficient to transport sand-sized sediments along the valley axes. Furthermore, turbidity flows with speeds rarely exceeding 3 km/hr are by no means uncommon events and in fact may occur every few days in localities where rivers introduce large quantities of sediment to the sea near the heads of submarine valleys.

Our early studies of currents in canyons offshore southern California may have failed to reveal the importance of these low-velocity turbidity currents, partly because here beaches are the chief source of sediment and there are no large (and few small) rivers entering near the canyon heads. Furthermore, the occurrence of turbidity currents would not ordinarily be recorded in many southern California canyons because great masses of kelp and sea grasses are carried downcanyon by the currents, entangling the instruments, stopping their operation, and frequently causing current meters to be swept away. However, in canyons and valley heads off large rivers (or off small rivers during flood conditions) with no kelp debris on the bottom, we found a very different story. In three out of four areas of this sort even our brief periods of current measurement yielded examples of slow turbidity currents, and in the fourth area, off the Fraser delta, the failure to record a turbidity current may have been because of the almost continuous dredging at the river mouth which removes great quantities of available sediment.

[1]Manuscript received, August 8, 1978; accepted, January 8, 1979.

[2]Geological Research Division, Scripps Institution of Oceanography, La Jolla, California 92093.

Although turbidity currents may be the major mechanism for transporting sediments down canyons and seavalleys, the ordinary, almost continuous currents also are important. Commonly, speeds of up to 30 cm/sec exist in these flows which we find alternately moving up and down valley floors. These currents can transport finer sands as well as great quantities of silt. Thus beach sand introduced by longshore currents may be permanently lost to the beaches due to these normal canyon currents. It also is important to know whether waste products such as the heavier sludge from sewer outfalls, which is locally dumped into nearshore canyon heads, will be carried continuously down the canyons into deeper water. Our study of canyon currents provides at least some of the answers to this problem.

Our current-meter records also provide oceanographers with information which was not previously available. We found that currents flow almost continuously up and down the canyon axes, in cycles that have a length that is related to both depth of water and range of tide. High frequency of reversals-in-flow-direction occurs in most shallow canyon heads, whereas in deep water, reversals-in-flow have a frequency closely related to the semidiurnal tidal cycles. The depth where currents become tidally related depends on the range of tide, being at shallow depth with large tidal ranges and at great depth with a small range.

Internal waves were found to advance along the canyons and other types of valleys, most commonly progressing landward into shallow water, but in some areas progressing into deeper water. The latter usually appear to be related to a source of water moving toward a submarine valley head, as for example where a valley is located off a large river mouth.

Much is still to be learned about these seafloor currents, and some of our information needs more documentation, such as our discovery that at least in one place the currents are essentially unidirectional, moving seaward as a countercurrent under the landward-moving surface currents. We still are puzzled by finding that some records show the fastest currents in a direction normal to the canyon axes. Suggestions that surface winds may play a part in producing such anomalies require more investigation, including the comparison of currents at various heights above the valley floors and their transport potential.

It is surprising to learn how little investigation of bottom currents has been done by geologists and physical oceanographers during the otherwise rapid growth of oceanography in post-World War II years. To petroleum geologists, eager to understand processes of sedimentation, this study would seem to be of paramount importance.

Part 1
Introduction

The measurement of currents in submarine canyons has been a research project at Scripps Institution of Oceanography since 1968. Previously, despite rather extensive attempts to develop current meters for measuring currents on the ocean floor, there were few published results. Brief notes were issued relative to currents measured in a few places on the deep seafloor, but almost nothing had come from the continental margins despite extensive study of ocean currents well above the bottom.

The first few current measurements in submarine canyons were made by Stetson (1936) in the Georges Bank canyons. These were followed in 1938 and 1939 by a few more measurements off California (Shepard et al, 1939). These used the primitive Ekman current meters that gave only an integrated current velocity for short periods, followed by long intervals before repetition could be made. For 30 years after this early work, there were virtually no measurements made either in submarine canyons or on any other part of the continental terrace.

In the recent operations, reported here, we used the Isaacs-Schick Savonius rotor-free vehicle current meters to obtain continuous records of currents for periods up to one month. Initially, we operated in the canyons off La Jolla, California, but this was extended to other areas, partly with the cooperation of other institutions and agencies. Although the first measurements were exclusively in the canyon type of submarine valleys (for types of marine valleys, see Shepard, 1965), we now include fault valleys, delta-front valleys, and fan valleys in our studies. We recorded most of our data near the floor of the valleys at about 3 m above the bottom, high enough to avoid most kelp or sea grasses. However, we frequently suspended current meters on the same line at both 3 and 30 m above the bottom and occasionally at other heights from 2 to 150 m. Additionally, we placed two or more current meters separately in the axis of a valley at different depths to obtain simultaneous records of the flows. We kept current meters in canyons for periods of time up to one month, but most of our records have a length of the order of four to seven days.

The total number of hours (of measurements) to date is approximately 25,000. These were made in 25 canyons and other types of seavalleys. The depths of measurement were from 46 to 4,200 m.

The areas of operation (Plate I) now include nine California submarine canyons from La Jolla to Monterey Bay, a fault valley 80 km off San Diego, two canyons off the southern end of Baja California, two canyons or seavalleys off the Rio Balsas delta in western Mexico, a delta-front valley off the Fraser River of British Columbia, two canyons and two seavalleys off northern Kauai, five canyons off the East Coast of the United States extending from Georges Bank to Chesapeake Bay, the Rio de la Plata Canyon off northern Puerto Rico, two canyons off northern St. Croix in the Virgin Islands, two seavalleys off the Abra delta of northwest Luzon in the Philippines, and the Congo Canyon off western Africa.

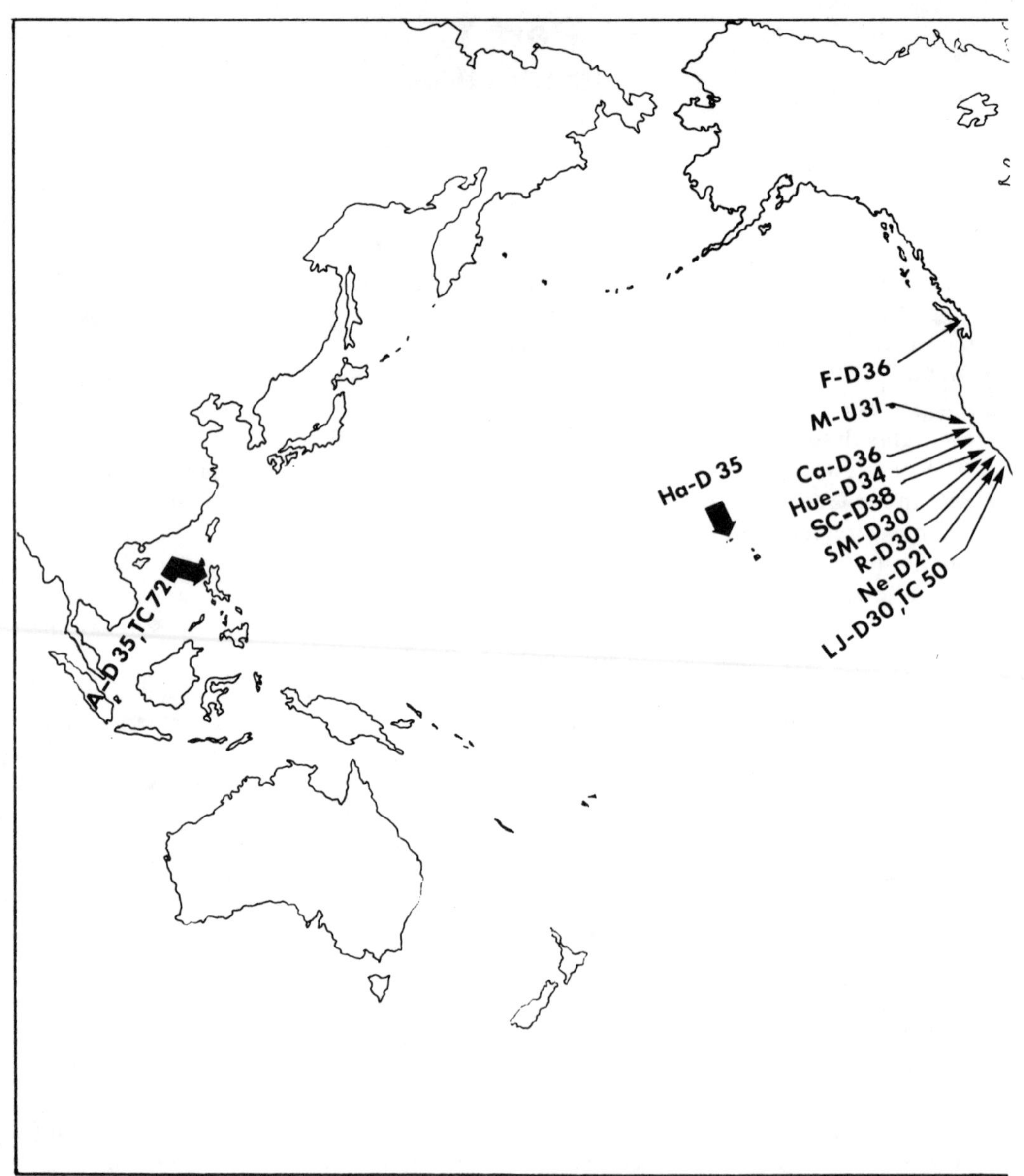

PLATE I–Index to areas of operation cited in this manuscript. These include areas off California, Baja, Mexico, Hawaii, The Philippines, British

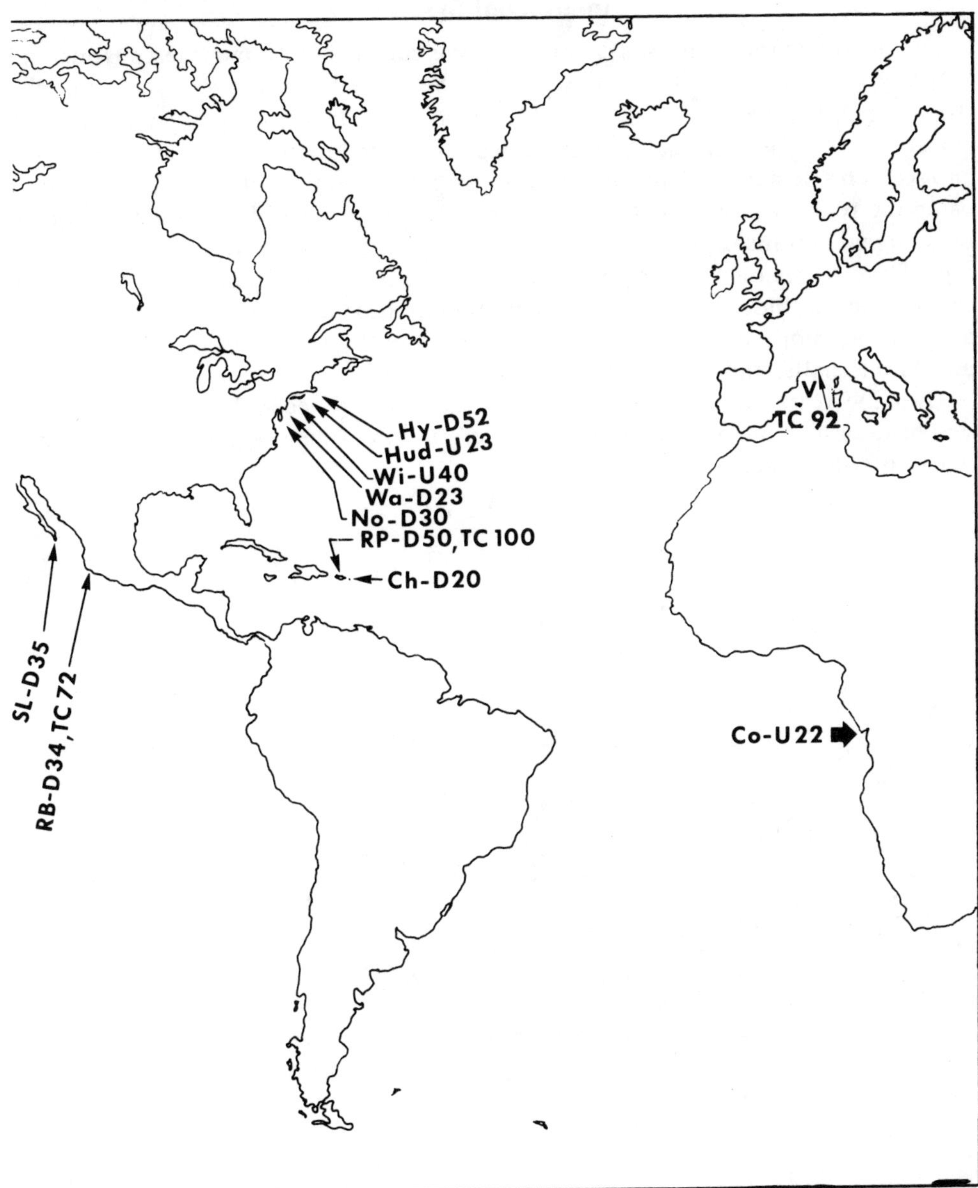

Columbia, U.S. Atlantic Coast, Puerto Rico, The Virgin Islands, Southern France, and The Congo.

Instrument System

The instrument system used to measure the currents in submarine canyons and other types of seavalleys was designed at Scripps Institution of Oceanography (Isaacs et al, 1966; Schick et al, 1968; Marshall, 1975). It consists of a Savonius rotor current meter (Fig. 1), and its unique feature is its low-threshold velocity of 0.5 cm/sec. This is achieved by optically sensing the rotation of the Savonius rotor using the reflection of a light source from a bright surface of the rotor. This light pulse signal is transmitted, by a photocell, to initiate a count which is recorded on tape. Drag forces on the rotor have been diminished by positioning the flat end of the tungsten carbide shaft on a spherical sapphire bearing, and by the neutral buoyancy of the rotor itself. The rotor is held in position by teflon bushings at the upper and lower shafts, and the system is water-lubricated.

Over the years our system has been updated by new technology. These changes will be described in conjunction with the description of the original system, much of which is still in use.

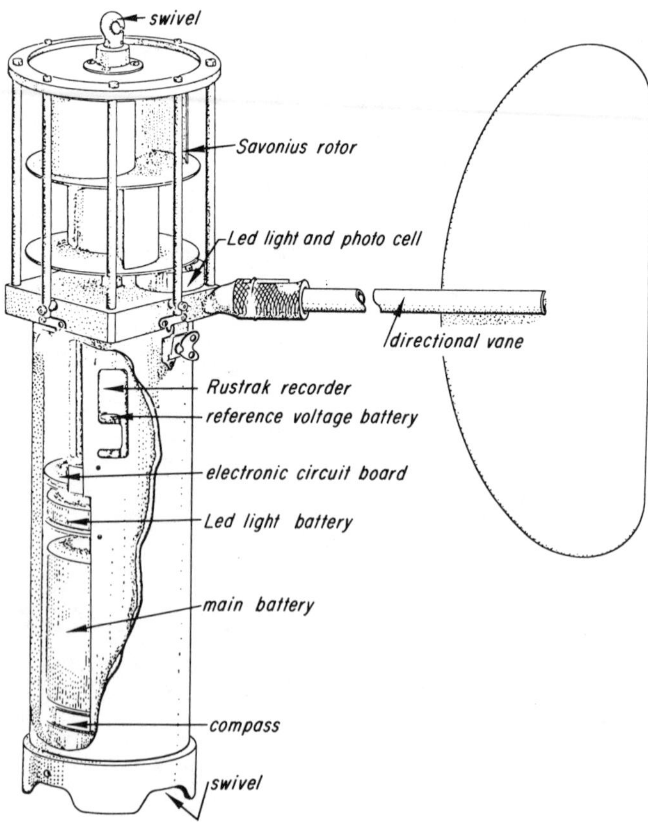

FIG. 1—Cut-away view of Isaacs Schick Savonius rotor current meter components.

Data acquired from the rotor are transferred to a Rustrak recorder as a tick mark, representing a preselected number of rotations of the Savonius rotor. The number of rotations per tick can be selected from a 2-4-8 geometric sequence to 256. At first we used mostly 64 rotations but after obtaining a quarter of our records, we used mostly 8 or 16. Combining this selectivity with the various available chart speeds of the recorder, it is possible to maximize the sensitivity of the system for conditions in any given area. Depending on the sampling rate selected, the maximum mission length is about 30 days. With the incorporation of new components, which will be described, it is now possible (within certain limits) to acquire one-year records.

Direction of current is obtained by permitting the entire swivel-mounted current meter to rotate from the impetus of currents against a fin mounted on a lever arm. The directions then are recorded in time sequence. Initially we used a fluid-damped, mechanical compass monitored with a variable resistance potentiometer. We changed to an eddy-current damped, magnetic compass with an optical sensor, but this was recently replaced by a flux-gate compass of a more compact and shock-proof design that has enhanced response characteristics. These changes have given us higher resolution and greater operational freedom.

We also incorporated into the basic system a cassette recorder which provides an ability to include records of temperature and pressure. The newer design permits basic field interpretation of data from the Rustrak recorder while eliminating the tedious digitization of records in the laboratory. The data transmitted to the cassette are recorded in preselected increments of 1-, 2-, 4-, 8-, or 16-minute periods as accumulated rotor speed counts. Thus the flexibility of having long-term missions or shorter-term, higher resolution data is provided.

The free-vehicle deployment package is a unique aspect of the system (Fig. 2). An instrument system, with weights and flotation, is released to sink to the ocean floor but keep the current meters constantly above the floor. After performing its function, it drops the weights (by the ignition of an explosive release) and returns the current meter to the surface using the incorporated flotation. Once deployed, there is no connection to the surface. This eliminates the loss of surface floats by pilfering or environmental hazard. The method also removes the effect of surface energy stresses in the record.

Where long-term deployment is desirable, electronic timed releases are used. The release system (Fig. 3) was designed at Scripps Institution of Oceanography by Sessions and Marshall (1971). It is a solid-state, programmable electronic clock with a total count time of 999 hours and an acccuracy of 5 sec/day. Recently an updated version was developed for maximum deployments of 9,999 hours or 416 days. When a pre-set time has been reached, a capacitor is triggered, setting off an electro-explosive pressure cartridge. The pressure rise propels a piston with an attached cutter, which separates a frangible section from the case, thus releasing the anchor weight. Although this release is reliable, we use a backup system to minimize loss due to occasional failure.

Methods of Presenting Results

Records obtained by the current meters can not be used directly to show the character of the currents. It is necessary to transfer them to various types of plots made partly with the use of computers to show the characteristics of the currents

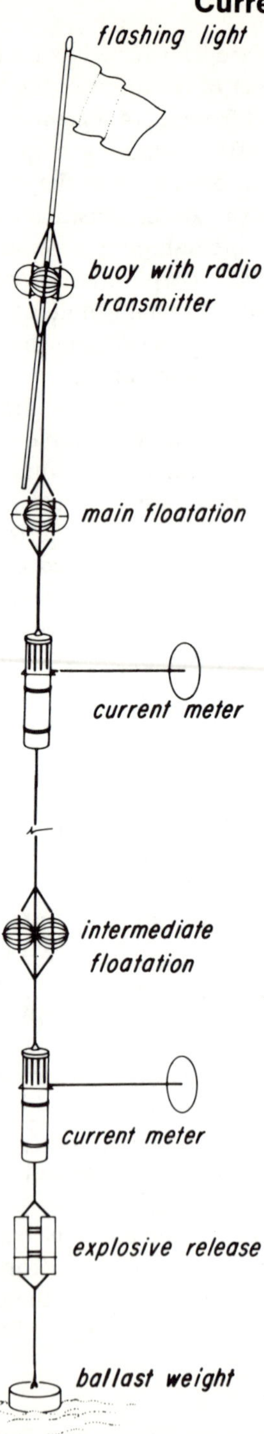

FIG. 2—Current meters deployed on the floor of a seavalley with release mechanism which allows dropping of ballast and a flotation device that returns meters to surface with buoy and flashing light.

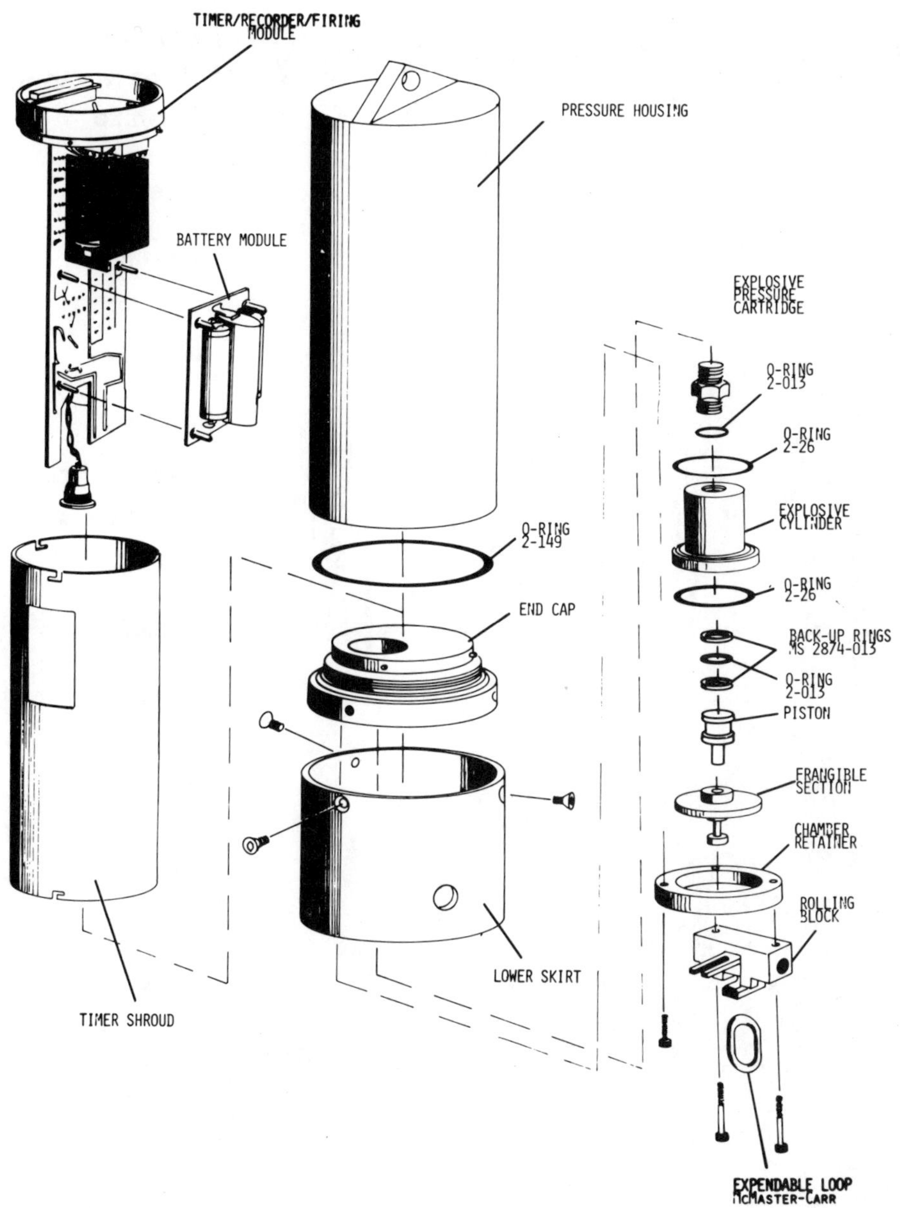

FIG. 3—Exploded view of the time release and components used in this study. Overall length is approximately 36 cm. A piston driven by expanding gases from the ignition of an explosive squib cuts loose a nipple from the frangible section. This permits a rolling block to turn, thereby releasing the ballast weights that hold the buoyant instrument package on the bottom.

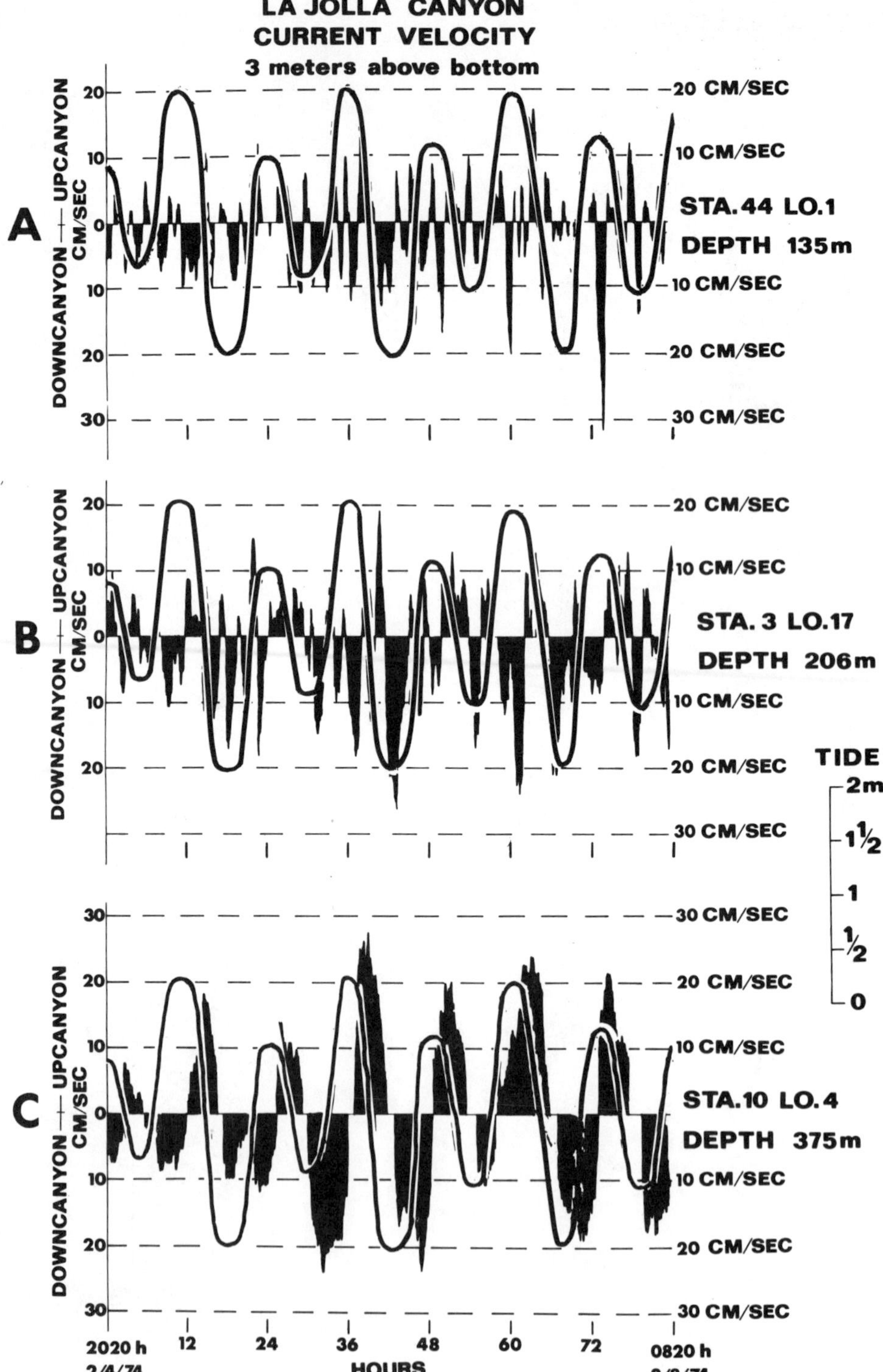

FIG. 4—Time-velocity curve of the currents in either the upcanyon or the downcanyon flows for unit periods in up- and downcanyon quadrants (vel plot). Tides obtained from tide tables are shown by the heavy curve. Alternations periods are longer at stations with greater depth and become related to tides at the 375-m station (C).

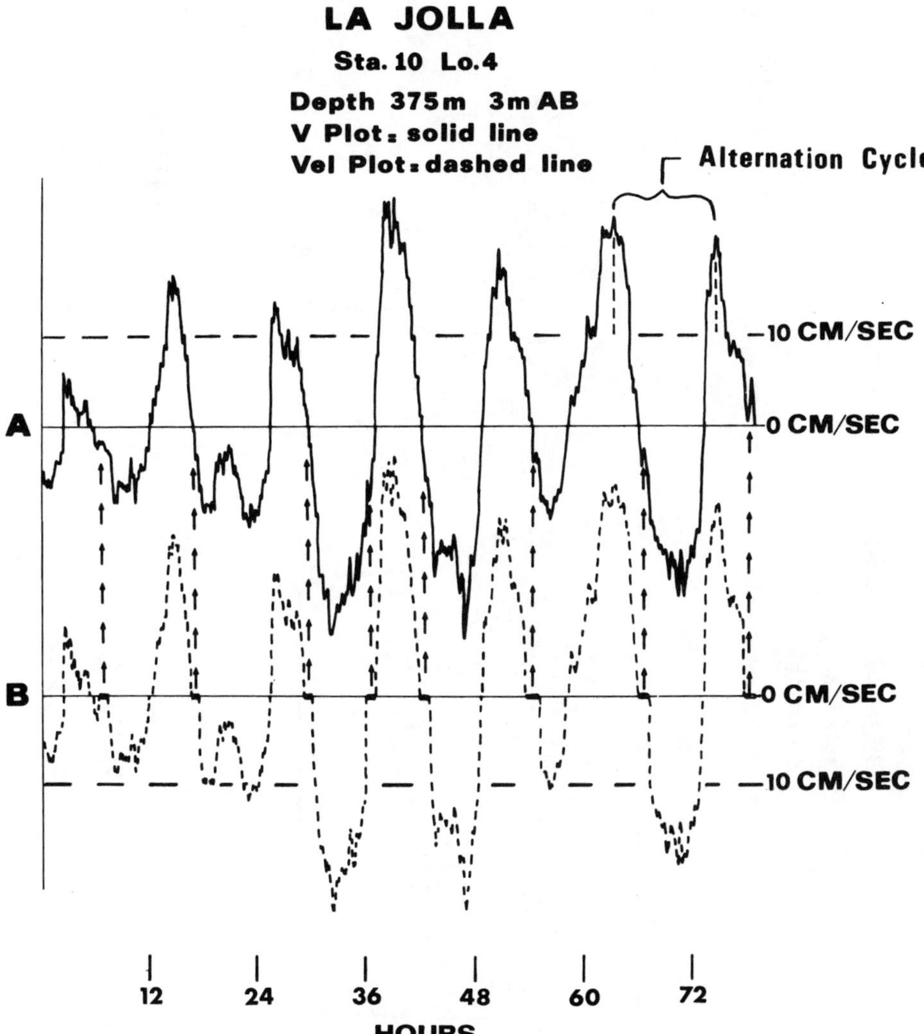

FIG. 5—Solid line shows the time-velocity curve, taken from the same record as Figure 4C, with speeds corrected by cosine of the angle between the axis and the flow direction (V plot). Only zero portions would be where angle is 90°. Dashed line shows the uncorrected speed plot with its zero portions wherever the current was more than 45° from the axis. Contrasts in the curves are emphasized by arrows. Alternation cycle is illustrated.

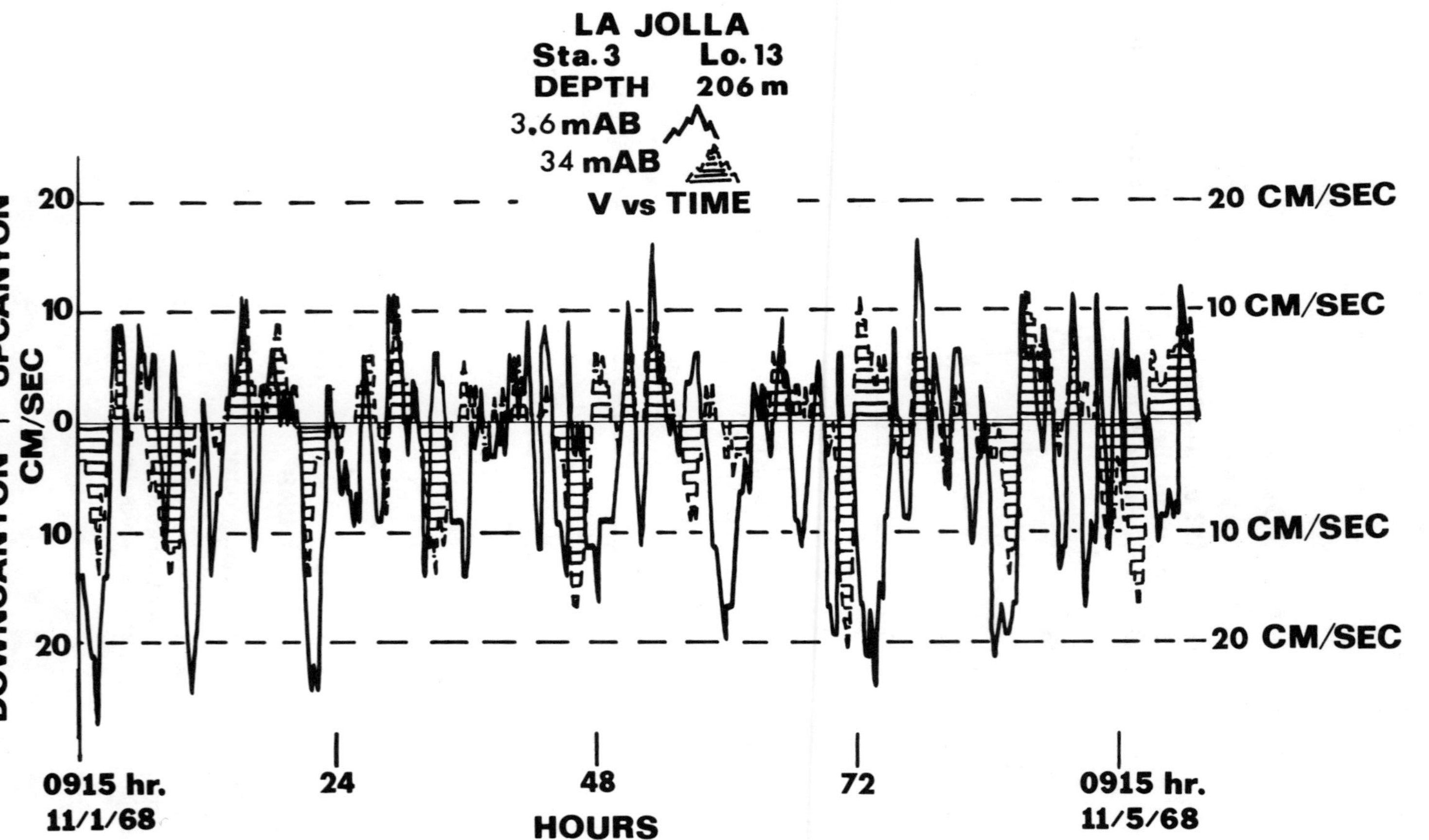

FIG. 6—Overlay of time-velocity curves at 3 and 30 m above bottom. Times of changing direction agree in much of the two curves although times of maximum flow generally are different. Considerably more agreement exists between the 3 and 30 m record at some other stations.

more graphically. The two most common methods are discussed here, and other methods are considered in Part III.

Because almost all of our records were obtained in valleys where the currents are predominantly along or close to the axis direction, it is best to convert the record tapes into vectors of up- and downvalley direction with speed given either as absolute speed of the up- and downvalley quadrants (called *vel plot* in diagrams, Fig. 4) or by multiplying the speed vectors by the cosine of the angle of divergence from the axial trend (solid line in Fig. 5, called *V plot* in diagrams). Both of these methods produce similar results and are referred to here as *time velocity curves.* They give a good representation of the currents as they move alternately up and down the valleys of the seafloor. The curves show how the currents vary with time both in direction along the valley axes and in speed. These plots also show what we call *alternation cycles,* indicating the time of an up- and downcanyon cycle.

Because we have many records which show the currents at two or more heights above the bottom, we developed a method to show how the speed and up- and downvalley direction vary at two different heights, although the method does not work effectively where more than two superposed current meters are used. Thus, in Figure 6 we show currents at 3.6 and 34 m above bottom, recorded simultaneously in La Jolla Canyon. A certain amount of agreement in the times of current reversals and in the speed of the currents will be observed here. In other similar diagrams we found both (mostly) good agreement at the two different heights and (occasionally) very poor agreement.

Part 2
Discussion

General

The extensive information concerning the currents in marine valleys that is included in Part III, Support Data, can be particularly useful in showing how much coastal marine sediment gets into deep troughs. The marine valleys investigated include predominantly submarine canyons which head near coasts where there is an abundance of sediment supply (such as off large river mouths or where longshore currents are transporting great volumes of sand along the shore and dumping it into the head of any submarine canyon that they intercept).

The study of currents in submarine canyons and other types of seavalleys from many parts of the world (Plate 1) has helped to solve problems concerning the origin of currents in these valleys. Our results would have been very limited had we studied only one area (such as the La Jolla submarine canyons or the Hudson Canyon) where it would have been easy to obtain oceanographic information. In Part III we describe the currents in numerous marine valleys. Here we can draw together the results and see what they tell us that bears on theories of sediment transport through valleys and into basins along the continental margins, and about the formation of submarine canyons and other types of marine valleys. Table 1 (at the end of Part II) includes the tabulation of all our current-meter operations including a few stations on the margins of the canyons.

The data and figures that are included in Part III will show that currents in submarine canyons and other types of marine valleys are much more comparable to tidal currents than to major ocean currents. Our data commonly show a reversal of direction at intervals comparable to those of the reversing tides, moving alternately up and down the valley axes with the period of reversal at intervals comparable to the reversal of the tides. However, much more frequent reversals are found in most shallow valley heads and even in relatively deep parts of those valleys where tides have a small range. These typical currents have a relatively slow velocity, rarely as much as a nautical mile an hour (50 cm/sec). However, they commonly attain velocities of 0.5 mi/hr (25 cm/sec), which is sufficient to transport considerable amounts of the sediment that we know is introduced into the heads of most marine valleys. The net flow of the currents in the valleys is more commonly down- than upvalley so that these currents are tending, with the help of gravity, to move sediments down the valley floors.

Probably of considerably more importance in the transport of sediments down the canyon axes are the unusual currents generally referred to as turbidity currents. Our records show four examples (Plate 1) of such currents, virtually the only complete records of this phenomenon that have been obtained. Although this might seem to indicate that turbidity currents are relatively uncommon, we believe we have established that relatively slow turbidity currents are actually very common in the valleys just off river mouths where deltas are building forward beyond the continental shelves. Because such situations probably were common in the past, it is at least probable that turbidity currents of this type were an important agent in supplying sediment to ancient basins.

Alternating Current Directions Related to Depth and Tidal Ranges

Even the earliest studies of currents in canyons (Stetson, 1936; Shepard et al, 1939) suggested that currents might alternate in direction between up- and down-canyon. However, it was thought from these early studies that the periods of alternation were not related to the tides. Nothing contradicting this idea was found before 1970 when we obtained our first records in water over 300 m deep. In these records we found direction alternations that were closely related to the semidiurnal tides (see, for example, Fig. 4C). On the other hand, the shallow stations, where the depth was less than about 250 m, continued to show alternations of direction of higher frequency[1] (see Figs. 4A,B, 42). In these canyons, all off California, tides have a range of the order of 2 m. We have taken records from areas with smaller tidal range, and have found that we had to go to much greater depths along the canyon axes before we found current-reversal cycles with a clear relation to the tides. Furthermore, we were led to suspect that, in areas with very small tides, alternations were more frequent than for locations with similar depths in areas with large tidal range (compare Figs. 4, 7). There is some indication that the cycles of direction alternations increase in length progressively until they become equivalent to the tidal period as in Figure 7. With results from 25 areas to compare, it is possible to show the relation of tide and depth to the cycle length, and how this relates to marine valleys in general. In Figure 8, most of the plots that show cycles of more than 10 hr are essentially tidal (12.4 hr) because there commonly are a few additional short-period current-direction reversals in addition to those which relate clearly to the tide (e.g., Fig. 9). Also where the cycle times are greater than the 12.4 hr for semidiurnal tides, there are direction alternations that clearly are diurnal, and the plot shows the average of all the alternations.

The plots in Figure 8 show an extraordinarily good relation between the axial depth at which the tidal relation develops and the range of the tide in various canyons and valleys. Thus at almost all of the California stations where the tides range from about 1.2 to 2.4 m, tidal correlation begins at axial depths of 250 to 400 m, whereas all areas with very small tidal ranges have shorter than tidal periods of alternations out to depths of 1,000 m or more. If we look at the ends of the spectrum, we find Christiansted Canyon in the West Indies where the tides are mostly less than 0.3 m and the tidal relation appears only at the 2,525-m station; at the other extreme, in the Fraser Seavalley with its tides up to 4.6 m, the relation exists at 60 m with an 11.7-hr cycle, and actually most alternations coordinate well with the tide.

Perhaps the best relation of cycles to tides and depths is in Rio Balsas Canyon and its tributary, Petacalco Canyon. Here, with a tide of about 0.6 m at maximum, we have 6 stations, all of which show an increase in cycle length with depth, and only the deepest (at 1,904 m) is clearly tidal (Fig. 7).

On the other hand, two of the canyons seem to show little relation of their alternations to either tide range or depth. Monterey Canyon (see Part III) has exceptional currents in almost every respect. It has few direction alternations that correspond with tidal periods even at the greatest depths. Santa Cruz Canyon is almost as

[1]Recent records during a Perigean tide show a fine tidal sequence in shallow water at the head of La Jolla Canyon.

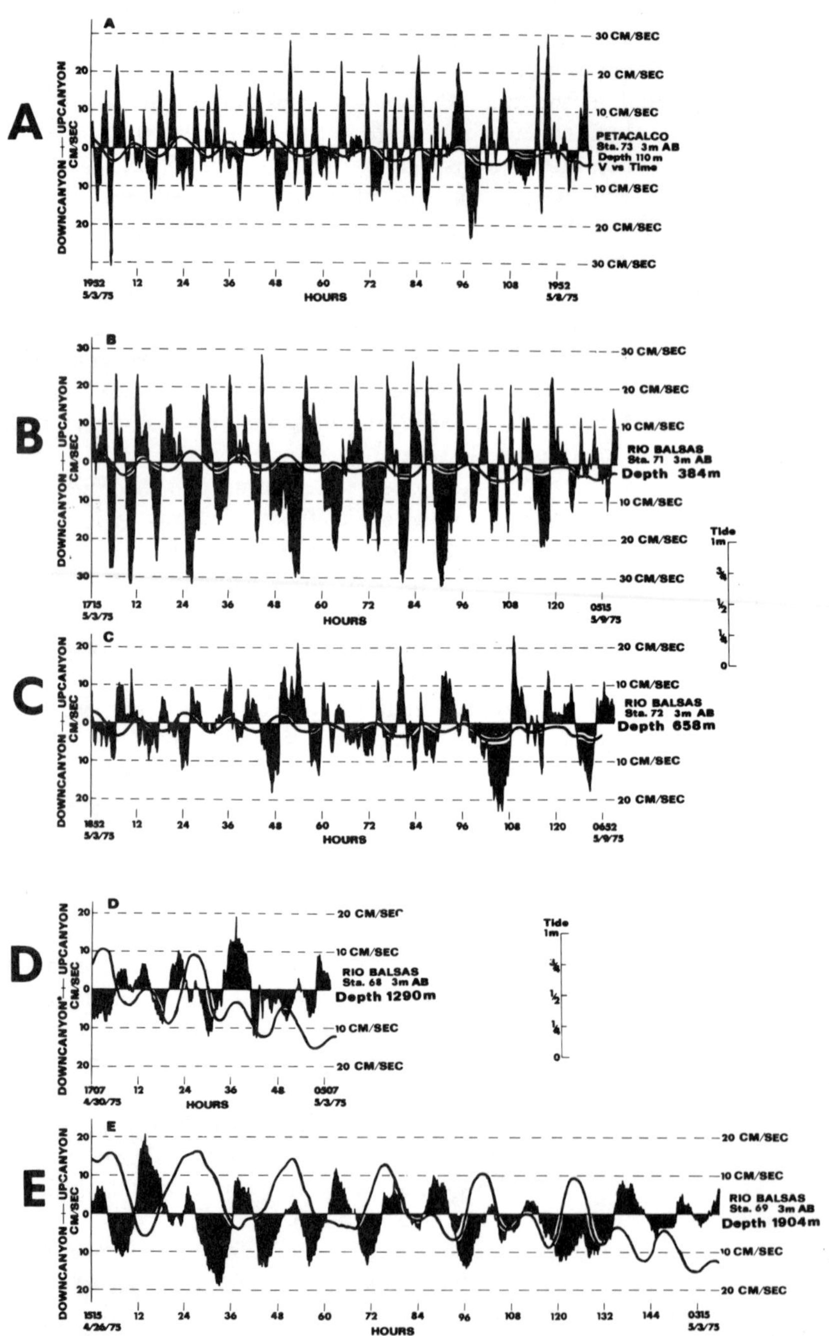

FIG. 7—Time-velocity curves showing the increasing length of alternating cycles with depth in the canyons off the Rio Balsas delta. The tides decrease in height because mud and sand covered gauge during operation.

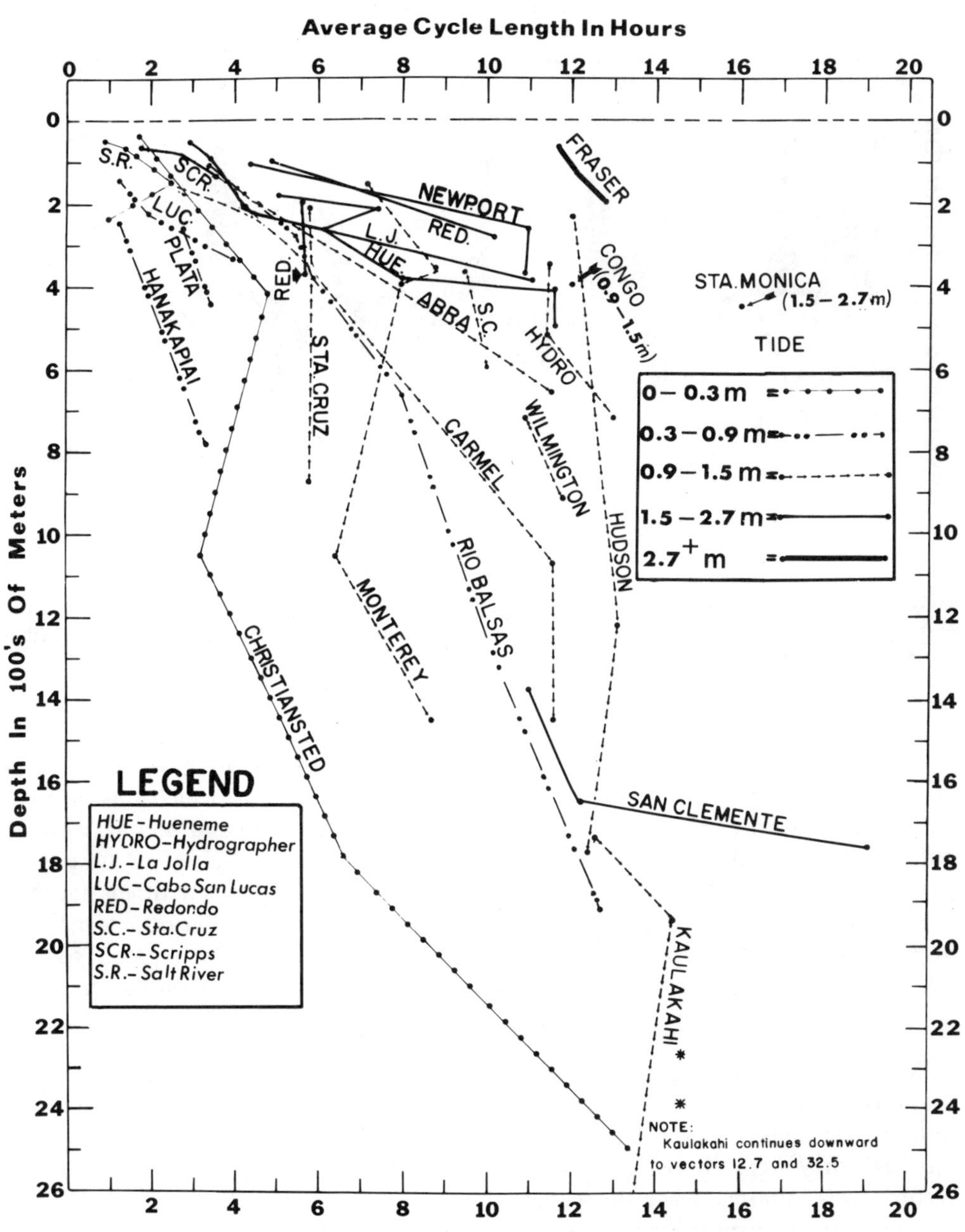

FIG. 8—Composite diagram showing relation of average cycle period of the up- and down-valley alternations compared with depth of valley axis and range of tide (see box on right). Lines showing nature of tide connect stations in the same canyon. Small tidal ranges and small depths are almost always associated with short average cycles; large tidal ranges and/or deep water are almost always associated with long average cycles.

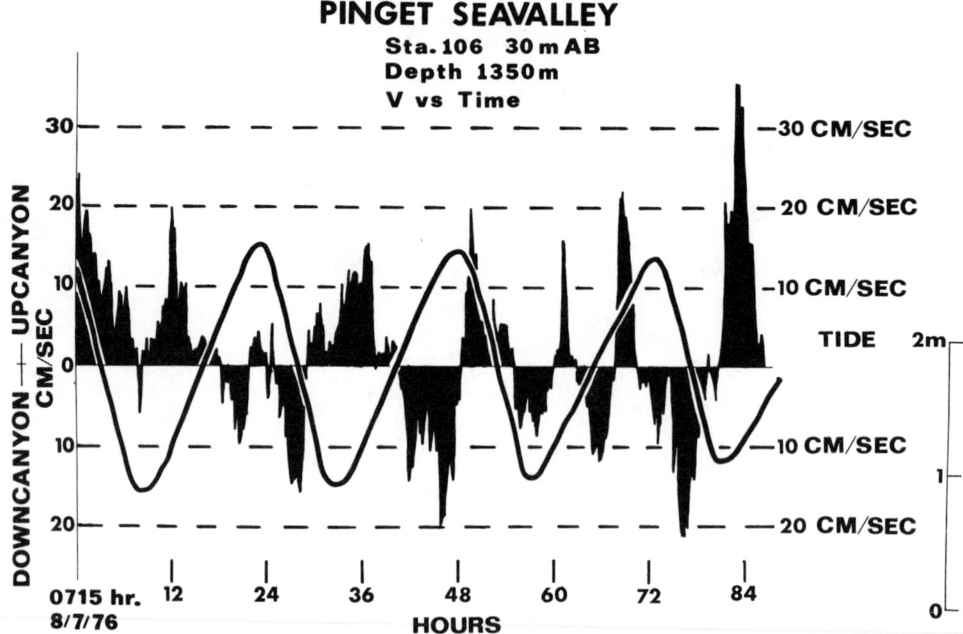

FIG. 9—Time-velocity curve in Pinget Seavalley at a depth of 1,350 m, 30 m above bottom (no 3-m record). Tide is diurnal, whereas the current flow direction alternations are rather close to the semidiurnal period particularly for upcanyon flows.

peculiar in relation to tides and depth. On the two occasions when we recorded in this canyon, the cycles were about the same at the two stations, which was particularly strange on the second occasion because the depths, 208 and 860 m, were so different. At both these canyons the tide has a range of 1.2 or 1.5 m, so there is no apparent reason for this failure to follow the pattern of most of the other stations.

We suspect that, in areas with small tidal ranges, where short-period alternations occur to considerable depths, the tide becomes subordinate to other regional oscillation frequencies which take over and exert their influence on canyon currents.

Records with Unidirectional Flow

We have no records that show continuous downvalley or continuous upvalley flows. However, in the head of Salt River Canyon at axial depth of 48 m, 3 m above bottom, we have one record that indicates almost entirely downvalley currents, although they were almost entirely upvalley at 30 m above bottom (Fig. 10). The records were obtained when the water column was isothermal, a very unusual condition in canyons even in depths as small as 48 m. Another place where the current was predominantly unidirectional was at 60 m in the Fraser Seavalley off the mouth of the Fraser River during our early June records of 1977 (Fig. 11). Here the current alternates largely at tidal intervals with very short, weak downvalley flows contrasting with stronger, longer continued upvalley flows. The apparent explanation is that a saltwater wedge moves shoreward under the seaward flow of low-salinity water introduced by the Fraser River.

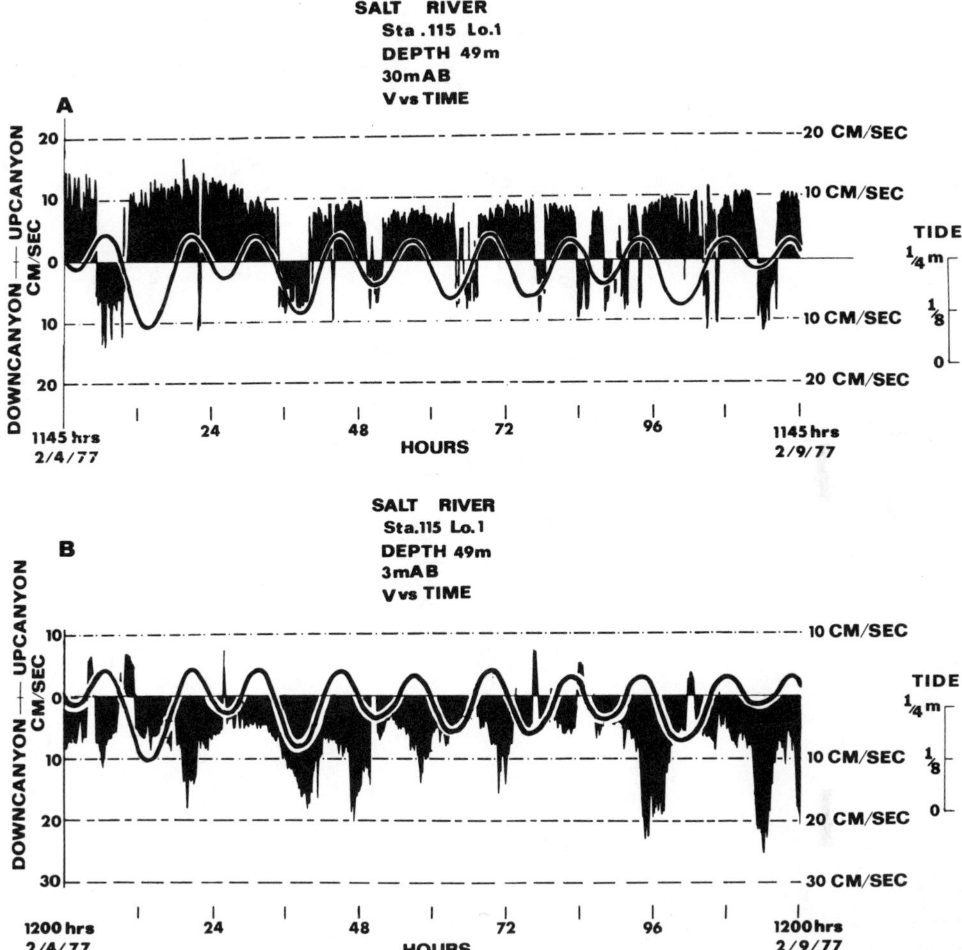

FIG. 10—Current meter time-velocity curves at 3 and 30 m above bottom at axial depth of 49 m in Salt River Canyon during isothermal conditions. Unusually strong trade winds are thought to have moved the surface water landward and a return flow occurred along the bottom out at least to this depth causing the almost constant seaward flow shown in **B**.

Among the few records of canyon currents obtained by others, one from the Var Canyon off southern France (Gennesseaux et al, 1971) during a storm and flood period, is of particular interest (Fig. 12). Apparently this shows nothing but downcanyon currents with strong pulses up to 90 cm/sec alternating with weak flows of about 15 cm/sec in the same direction. Like the example in the Salt River Canyon, Var Canyon is in an area of small tidal range which may be significant, applying as it does to both examples. There is no information that tells us whether or not the water is isothermal in Var Canyon. In any case, the high-speed flows in Var Canyon are probably the result of turbidity currents developed from the large volumes of sediment introduced by the Var River.

Another case of unidirectional flow was reported by Cacchione et al (1978) from a few days of observations in Hudson Canyon at a depth of about 3,000 m. The currents were weak and variable in speed. No explanation for this unidirectional flow is available.

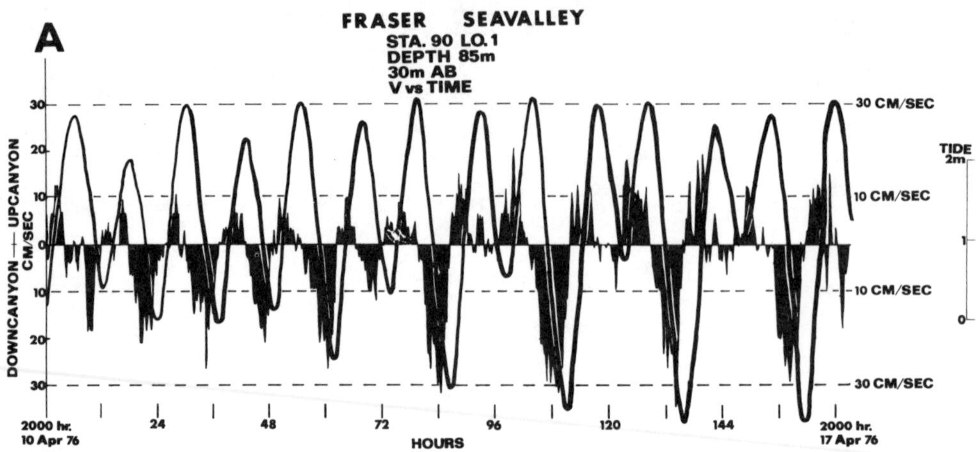

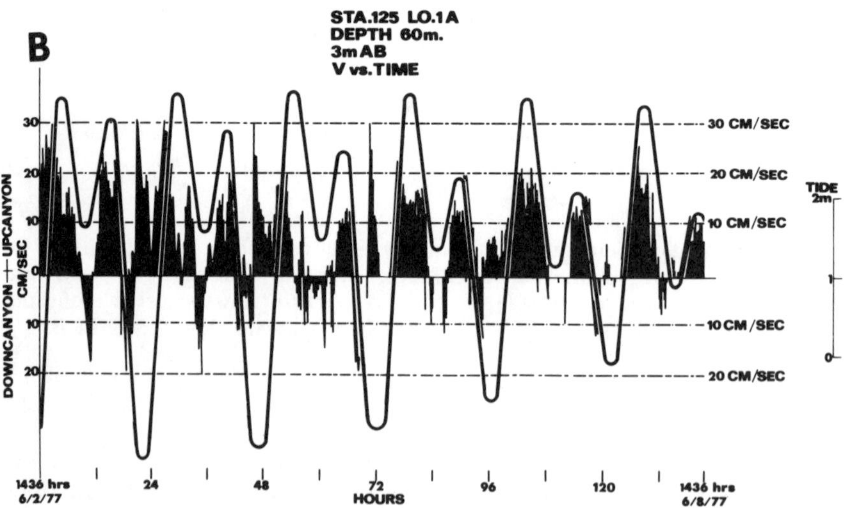

FIG. 11—Contrast between currents observed at the station near the head of Fraser Seavalley in 1976 (**A**), with those of 1977 (**B**). The two are not quite comparable, as the 1976 records were made at 30 m above bottom and the 1977 at 3 m. Both records show current alternations in essential agreement with the tidal periods although the 1976 currents are more related to diurnal tides for the last half of the record and all of the 1977 record is related to the semidiurnal tide. The dominant flow was downvalley in 1976 and predominantly upvalley in 1977. Station 90 showed excellent agreement with tide phase.

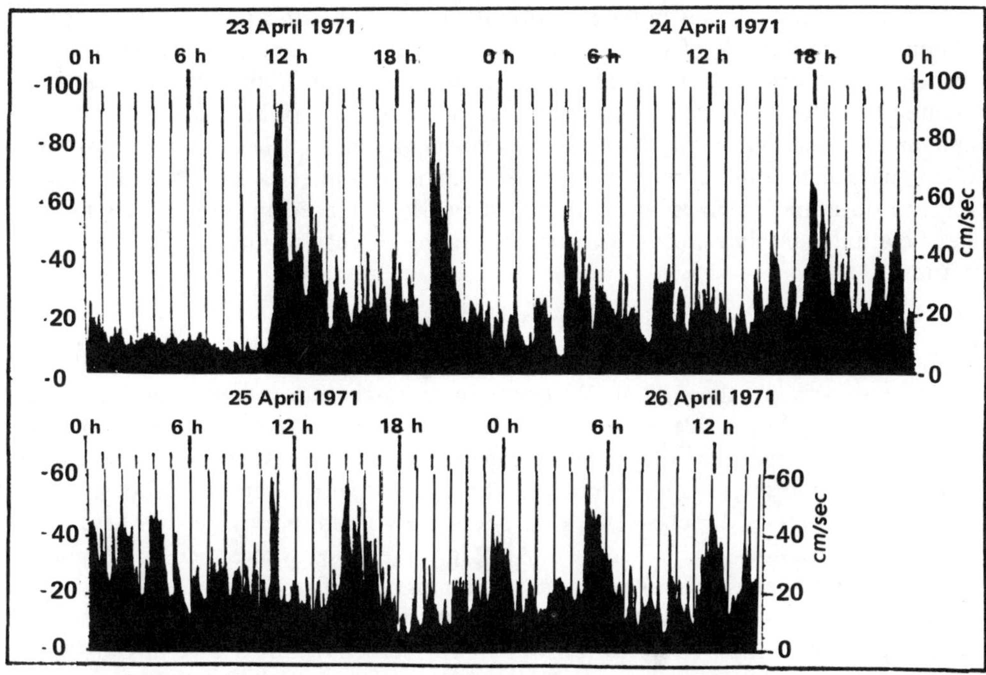

FIG. 12—Turbidity currents in Var Canyon off Var River during flood and storm conditions in southern France. Currents are unidirectional downcanyon and show pulses of strong flow at approximately 8-hour intervals (from Gennesseaux et al, 1971).

Net Flow in Canyons and Seavalleys

It is important to determine whether net flow in canyons that extend in close to the coast is up or down the axis, as is shown in our progressive vector diagrams. Sewer outfalls may be located near canyon heads, and at such places flows can carry the sludge downcanyon whereas net upcanyon flow will tend to carry much of it along the shelf (with a good chance that it may be returned to the shore and cause pollution).

Our investigations rarely are sufficiently frequent to provide adequate opinions on net-flow directions. However, we plotted 69 examples at 3 m above bottom where we found a net flow either up or down the axis. Of these, 43 showed a net downcanyon direction, 26 showed net upcanyon, and 3 showed no appreciable net movement. There is little choice in the case of the currents measured near the canyon heads, as the up- and downcanyon flows nearly balance and, in La Jolla and Scripps Canyon, all of our many currents measured at less than 150 m showed net upcanyon flow. Upcanyon net flow seems to be more common in the East Coast canyons where there are more net upcanyon flows than downcanyon, although heavy mineral studies, ripple marks, and bottom photos (Keller and Shepard, 1978) indicate the ultimate net flow may be downcanyon.

Conversely, most canyon measurements along the California coast indicate downcanyon net flow as did most of the records from canyons of the Cape San Lucas area off Baja California and from Rio Balsas Canyon farther south, off western Mexico.

Our measurements mostly reflect normal current flows and rarely include observations during periods of high storm energies. Because numerous studies indicate dispersion of sediments seaward and not landward in canyon systems, we must assume that during storms great quantities of material may be transported downcanyon and that net sediment transport is downcanyon. Our data may be somewhat inadequate as to long-term net flow because many of our recordings are only 3 to 5 days in duration. In cases where flows are nearly balanced in both directions, the net flow may simply reflect the fortuitous timing of current-meter installation and retrieval.

Consistency of Speed

Tides coming into a bay ordinarily show a gradual increase in speed until they reach a maximum, then gradually decrease to slack tide followed by a gradual buildup of the ebbing tide to maximum (Fig. 13). Such a sequence is uncommon in the currents of marine valleys. More commonly, after a reversal of direction the currents build to their highest speed in a short period with several peak flows before the current reverses direction (see Fig. 14). A slow buildup may occur where there are tidal alternations, as in Figure 68. Also, in some cases, where speed is very slow, periods with no measurable velocity are common (as in Fig. 30 where we see a large number of zero velocities during the 15-minute data sampling intervals).

The pattern of rapid acceleration was first observed during a dive in Cousteau's *Diving Saucer* in La Jolla Canyon (Shepard, 1965). The current was moving slowly downcanyon and suddenly stopped; then it reversed and built up so that in a few minutes ripple marks began to form with their steep side upcanyon. According to George Keller (personal commun.) a buildup from 2 or 3 cm/sec to 25 cm/sec occurred in a sinkhole 230 m deep off eastern Florida.

An exception to this tendency of rapid buildup of canyon currents developed in Salt River Canyon at 49 m (Fig. 10). There, an examination of the current-meter records showed an almost continuous series of rather equally spaced speed indications with no observable slackening of flow during the rather rare reversals.

Sediment Transport (Threshold Velocity) and Maximum Velocities of Ordinary Currents

The significance of currents in canyons and other seavalleys relates to the capacity of currents to reach threshold velocities necessary to transport sediment along canyon axes, and to their ability to erode sediment from the floors and walls of seafloor valleys. One of our problems is that most of our information comes from 3 m above the bottom largely because of the nature of the instrument (Fig. 2) and the danger of fouling or interference by bottom debris. We have little to indicate whether currents at the sediment surface are comparable to those at this height above the floor. Sternberg (1965, 1967, 1971) and Sternberg and Creager (1965) developed a flow profile that extends upward to only 1 m above bottom. Miller et al (1977) further defined the boundary layer to 1 m above the bottom. On one occasion we obtained speeds at both 2 and 4 m above bottom. We are attempting to obtain measurements much nearer the sediment surface; meantime, we refer only to speeds obtained at 3 m above bottom.

Much has been written about the minimum (or threshold) velocity necessary for erosion and the transportation of various sizes of sediments (Shields, 1936; Hjul-

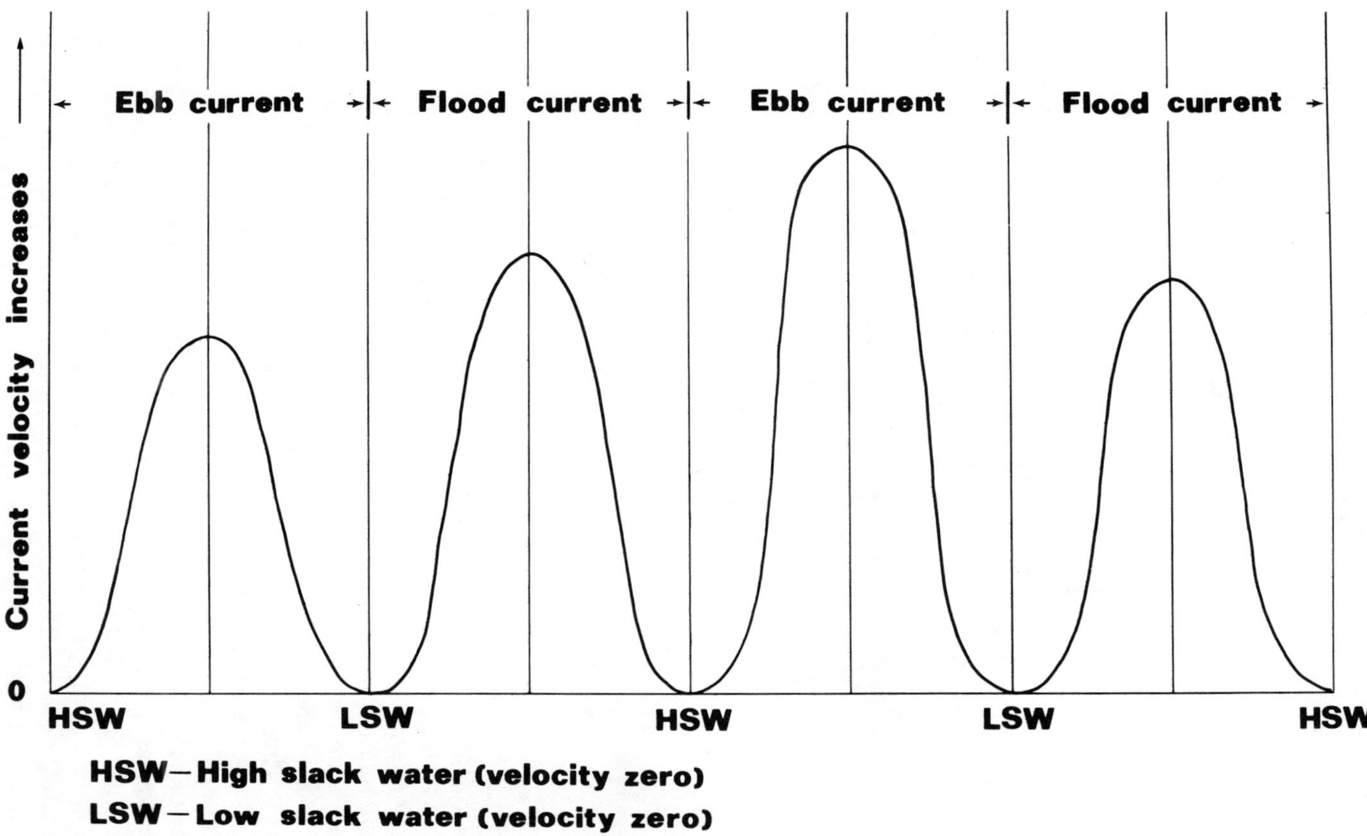

FIG. 13—Typical increase and decrease of current speed as tide comes into a bay through a narrows and later leaves the bay. Note contrast with the speed development in up- and downcanyon currents.

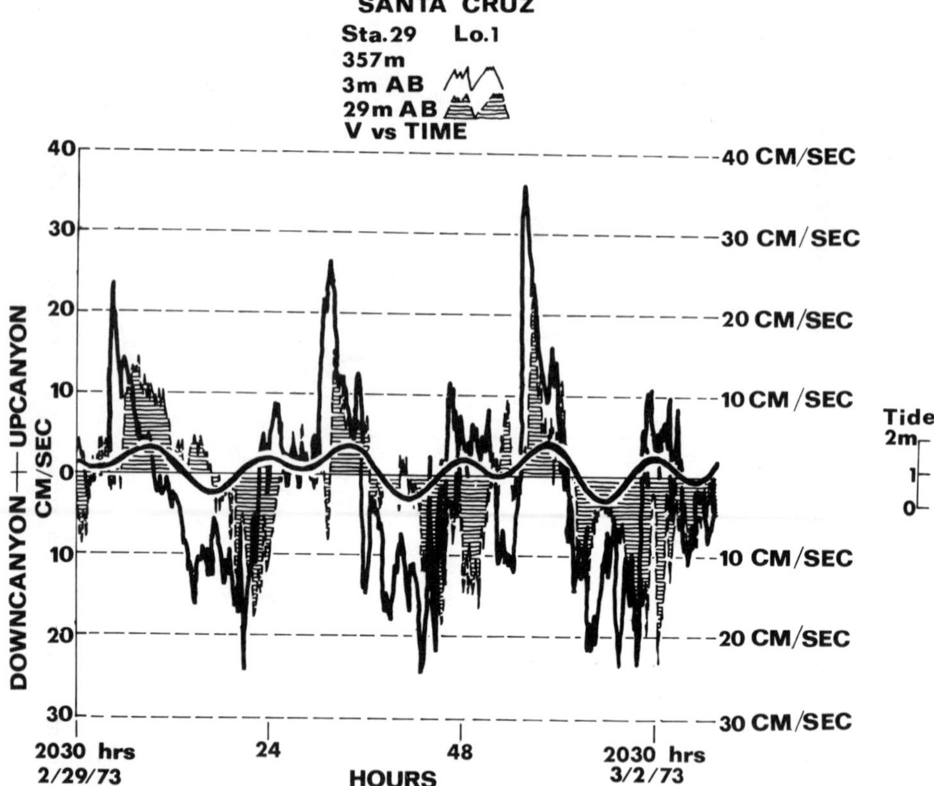

FIG. 14—Overlay of records taken at 3 and 29 m above bottom at a depth of 357 m in Santa Cruz Canyon. The same diurnal interval between up- and downcanyon peaks is observed but those at 29 m above bottom occurred somewhat later. The heavy line indicates the tide.

strom, 1939; Sternberg, 1965, 1967; Sternberg and Creager, 1965; and Miller et al, 1977). The complexity of these methods and their many contradictions were clearly documented by Gorsline and Swift (1977). It is almost impossible to apply this very valuable information to our current-meter measurements because we generally lack the vital information of sediment type and size, and have little knowledge of sediment on the bottom at the location. Therefore, we must rely on general observations by scuba divers, observers in deep-diving vehicles, and from time-lapse bottom photography. Thus, Keller (Keller and Shepard, 1978) reported observing sediment motion in Hudson Canyon from the submarine *Alvin* when the velocity reached 27 cm/sec. He used cameras for time-lapse photography at 710 m in the axis of Hydrographer Canyon from which successive photos showed ripple marks in the sand advancing downcanyon whenever the current (at 100 cm above bottom) exceeded 26 cm/sec. F. P. Shepard, diving in Cousteau's *Soucoupe,* observed the formation of ripple marks in the silty, fine-sand bottom of La Jolla Canyon, with a current of approximately 25 cm/sec. Data of this sort provide little to work with, but suggest that we can expect sediment movement, at least of fine mineral sands, where the currents are of the order of 25 to 35 cm/sec (0.5 km) at small heights above bottom. Ordinary currents in canyons reach these speeds commonly.

Because maximum currents of any particular period are the most important for sediment transport, we plotted the maximum speed versus the depth of water for both upcanyon and downcanyon for most of our records at 3 m above bottom[2] (Fig. 15). There is wide divergence of the maximum speed at shallow axial depths with a relatively low average for these maxima. Our shallowest station in Scripps and La Jolla Canyons showed consistently low currents. The speed increases with depth up to about 500 m and then declines slowly as the depth increases.

Our fastest currents, with exception of turbidity currents, are clearly in the United States East Coast canyons (Fig. 16) and in the Rio de la Plata Canyon north of Puerto Rico (Fig. 26). It is hard to say what accounts for the relatively high speeds of these currents in relation to other areas. The two highest readings were at the shallowest stations of their particular canyon, but this is by no means a rule for canyons in general (Fig. 17). In reassessing our records, we find that in canyons with the largest number of stations only one has any indication of decidedly faster currents at the canyon head and several show particularly slow currents at this location. Nor do the highest current speeds occur during strong winds (aside from occasional turbidity currents as observed off La Jolla). The fastest currents in Hydrographer Canyon occurred during relatively calm periods, whereas in Carmel Canyon, during a rather heavy blow, the currents barely exceeded 30 cm/sec. At times an increase in current speed actually occurred during decreasing wave and wind conditions. Nor are the high speeds related to large tide ranges. The Rio de la Plata area has a small tide range, and the East Coast canyons all have rather intermediate tide ranges.

Maximum speeds of upcanyon currents more commonly are slower than those of downcanyon currents (Table 2 at the end of Part II). There are few striking exceptions as in the mid-depth station in Hydrographer Canyon and at the two deepest stations of Monterey Canyon. The number of stations with downcanyon maxima is 89, with upcanyon maxima 44; 22 have the same velocity for both. That result is much less favorable to the thought expressed earlier in our studies that downcanyon flow is dominant (Shepard et al, 1974). Also, the deepest stations shown in Table 2 all have maxima in upcanyon direction or have the same speed for both. The deepest station of all (at 4,205 m) apparently has about the same speed for up and down so far as can be discerned from the very low velocities.

We conclude that these normal currents are capable of redistributing sediments along the canyon axes except perhaps at great depths. Some normal currents also are capable of producing features such as ripple marks in silts and finer sands. Because the strongest currents are more often downcanyon, these ripple marks will have steep slopes downcanyon and, with few exceptions, this is what has been found both from deep-diving vehicles and from compass-oriented photographs (Shepard and Dill, 1966). Finally, it should be recalled that we have very few records during storms.

Relation of Currents to Height Above Bottom

In Part III we show how the currents 3 m above bottom of the valley axes compare with those 30 m above bottom and at other heights above the valley floors. Here again some of our early impressions failed to provide the complete picture; in

[2]Repetitive stations off La Jolla could not all be included because of space.

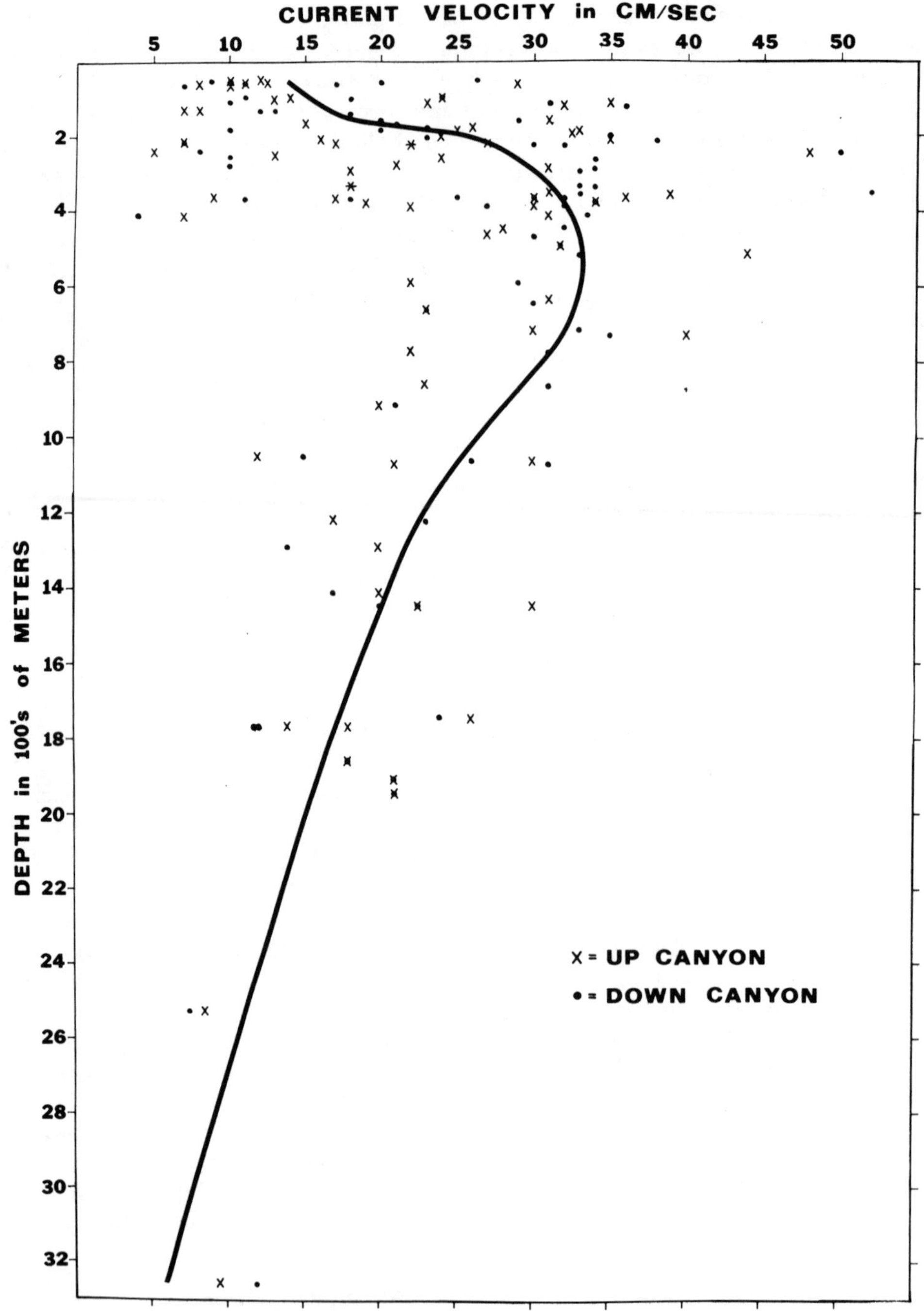

FIG. 15—Maximum speed of upvalley and downvalley flows at 3 m above bottom plotted against depth of valley axes. The speeds at depth too great to plot were less than any of those shown. Line shows general trend.

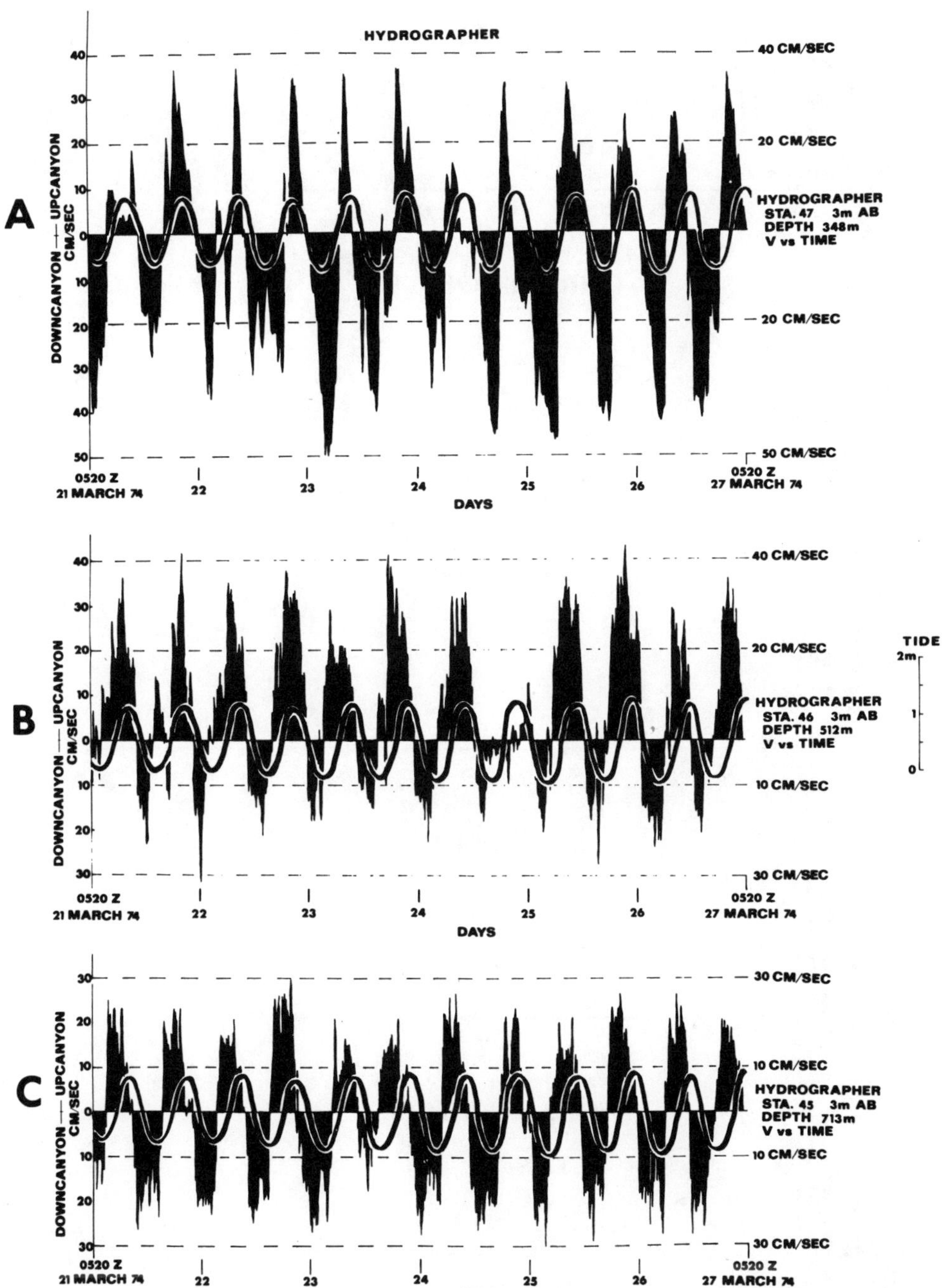

FIG. 16—Portion of time-velocity curves of contemporaneous records from Hydrographer Canyon. Tide curves show a rather consistent, progressive change in position in relation to increasing depth in the canyon, indicating that internal waves are moving up this canyon and accounting for the character of the current alternations. Note the excellent relation of currents to the tides. Probably the tides have a larger range than those on Georges Bank used here.

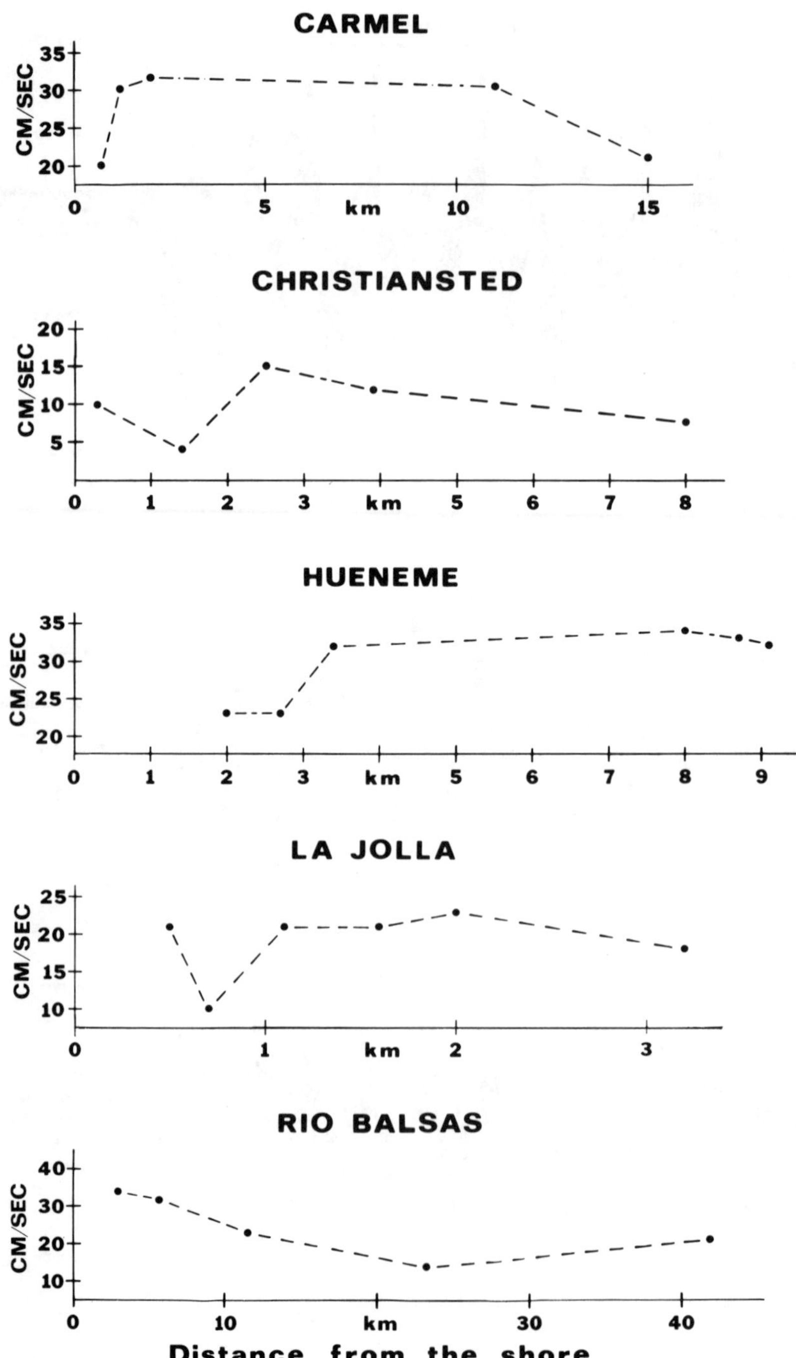

FIG. 17—Variation in the maximum downcanyon speed at stations versus their distance from the canyon head (equivalent to the shore in these cases) is shown. Canyons were chosen which had the largest number of stations.

fact, we still need more information about some aspects of this relation. Our first comparisons in La Jolla Canyon showed only rough agreement between currents at the two heights (Fig. 6). We later found (at several stations in Carmel Canyon) that the current measurements, first at 3 and 19 m above bottom and later at 3 and 30 m above bottom, agreed remarkably well (Fig. 18).

However, there was little agreement of current speed and direction between the 3- and 30-m heights at most very shallow-water stations, as for example at 49 m in Salt River Canyon, where the 30-m-above-bottom meter would be very near the surface as well as near the level of the canyon lip, and in Petacalco Canyon at 110 m where the flows often were in opposite directions (Fig. 19B).

The agreement between 3 m and 30 m in Carmel Canyon at 348 m depth is interesting because the upcanyon directions shown in the polar plot (Fig. 58) are almost at right angles to the downcanyon flows at 3 m.

In fact, the exact direction at 30 m above bottom is commonly different from that at 3 m (see Figs. 55, 81B, D). Mostly this can be explained as the result of less confinement of the currents by a deep, narrow gorge below the higher meter.

In general, there is much more than accidental agreement in the currents at the two heights above bottom, suggesting that they are moving somewhat as a unit up and down the canyons to levels of at least 30 m above the bottom. They probably are not influenced by the type of boundary-flow eddies that one expects in river flows such as in the deep water of the lower Mississippi.

Turbidity Currents

Since Daly (1936) first suggested that submarine canyons were caused by turbidity currents, and Kuenen's extensive tank experiments (1937) and his field work on turbidites did so much to interest geologists in these currents, there has been widespread opinion among geologists that occasional turbidity currents with high velocities, perhaps reaching even 90 km/hour (Heezen and Ewing, 1952) have cut submarine canyons. Theorists have also predicted high speeds such as 24 m/sec (Inman, 1963). With our current meters, we can not obtain records of such high-speed currents, but we have had current meters carried away during two storms at La Jolla, California, just as happened to Inman et al (1976) during another storm period in which a current velocity of 190 cm/sec was recorded at the Scripps Institution of Oceanography pier before the current meter was lost. We assume that a velocity able to carry off a meter is at least that high.

Our first solid evidence about turbidity currents came in December 1972 when we placed two superposed current meters in La Jolla Canyon just prior to a storm, and the current meters were carried downcanyon by a rapid flow. They were subsequently found and retrieved with the use of the submersible *Nekton Gamma* because the float ropes attached to them were seen extending above a tangle of kelp in which they were lodged. The instruments were undamaged by their 0.5-km voyage downcanyon. Therefore it seems likely that this turbidity current was not of a violent nature. As explained in Part III, the record shows that one downcanyon pulse reaching a maximum of 50 cm/sec (Fig. 20) preceded the transport of the instruments, but presumably a stronger flow followed (Shepard and Marshall, 1973).

During the submersible dive to retrieve the two current meters displaced during the storm, a noticeable cut had been eroded into the canyon floor deposits (Shepard and Marshall, 1973). Incorporated within the deposit and subsequently exposed by

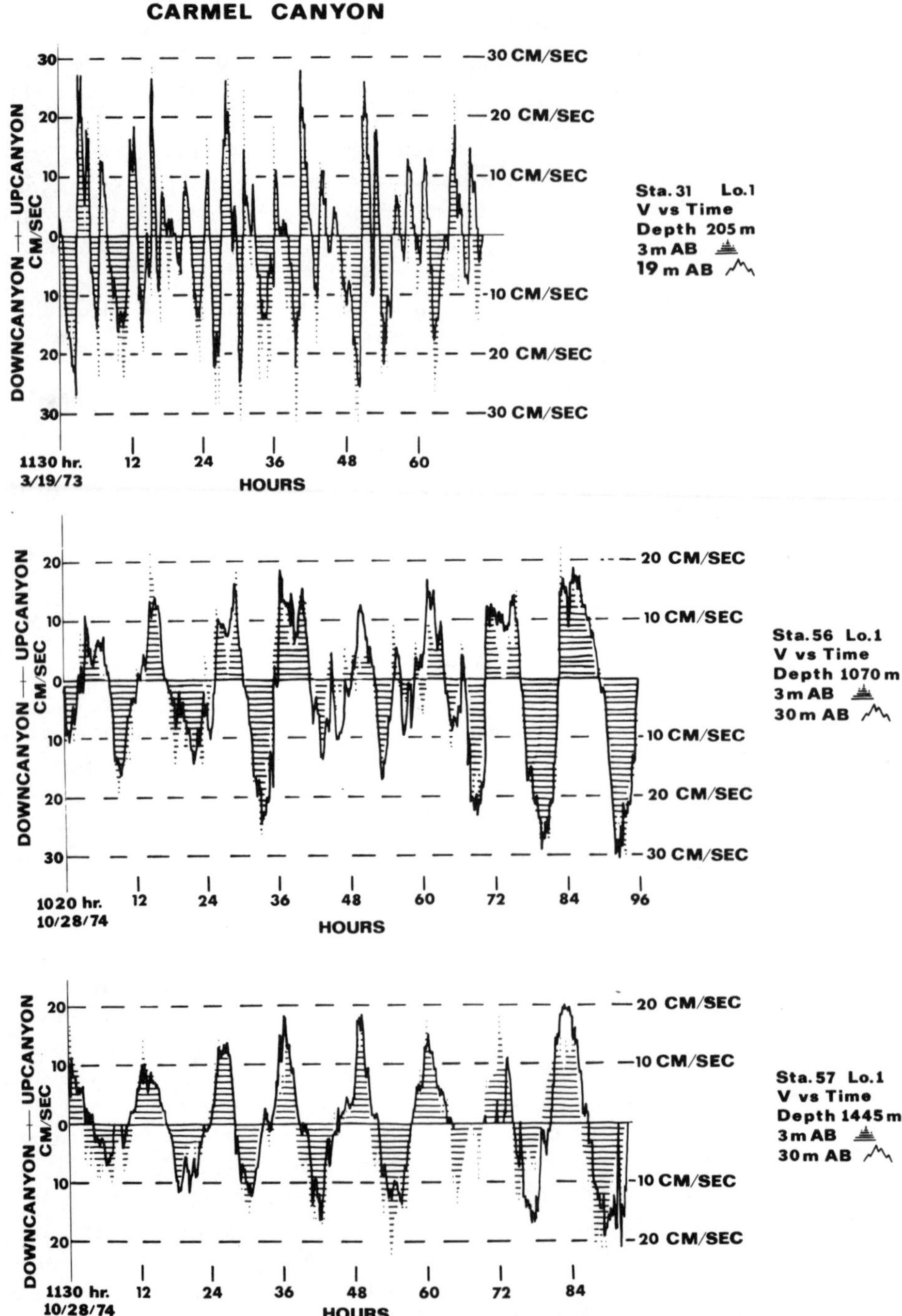

FIG. 18—Figure shows the good agreement between current meters at 3 and 19 m and at 3 and 30 m above bottom at various stations in Carmel Canyon. Equally good agreement was found between 3 and 30 m at a depth of 348 m (Fig. 58). In no other canyon was there such a good agreement at the two heights above the bottom.

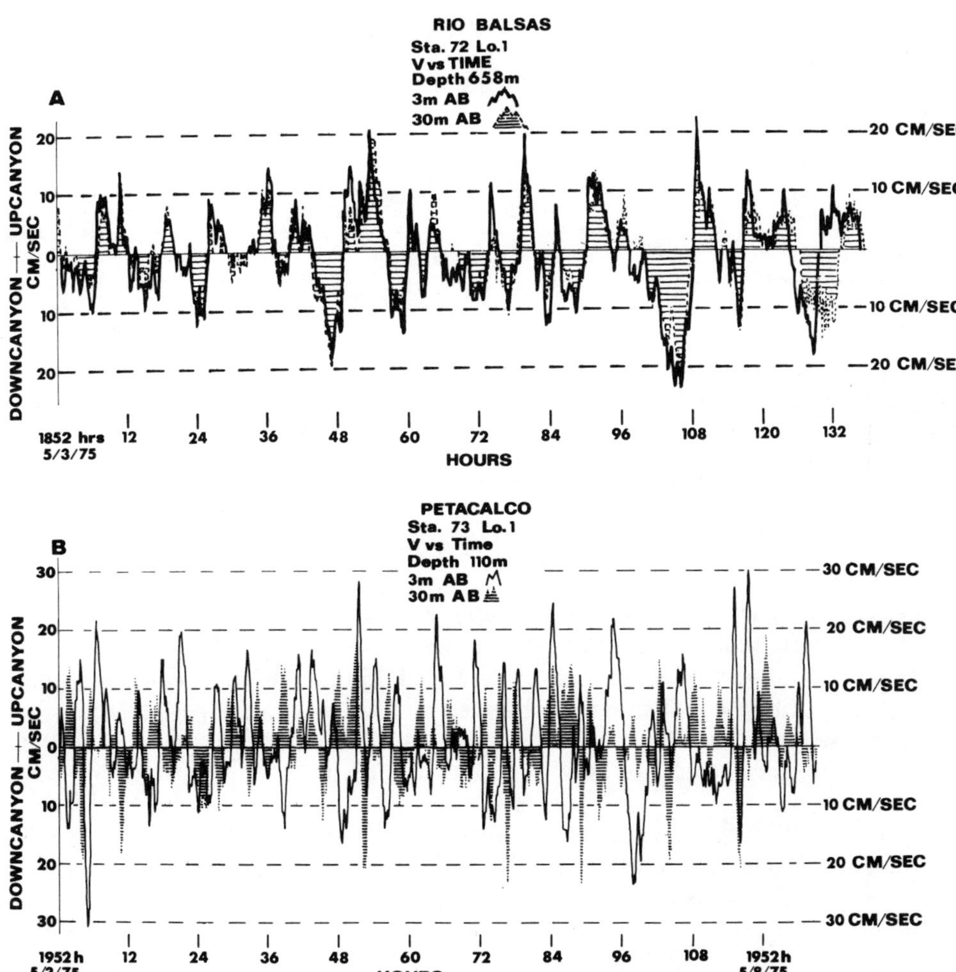

FIG. 19—**A**, figure shows the overlay of 3 and 30-m time-velocity curves at 685-m depth in Rio Balsas Canyon. Note the rather close agreement of times of direction change and speed of currents. **B**, an overlay of time-velocity curves at 3 and 30 m above bottom in Petacalco Canyon at 110-m depth. Note that currents at the two heights often flowed in opposite directions.

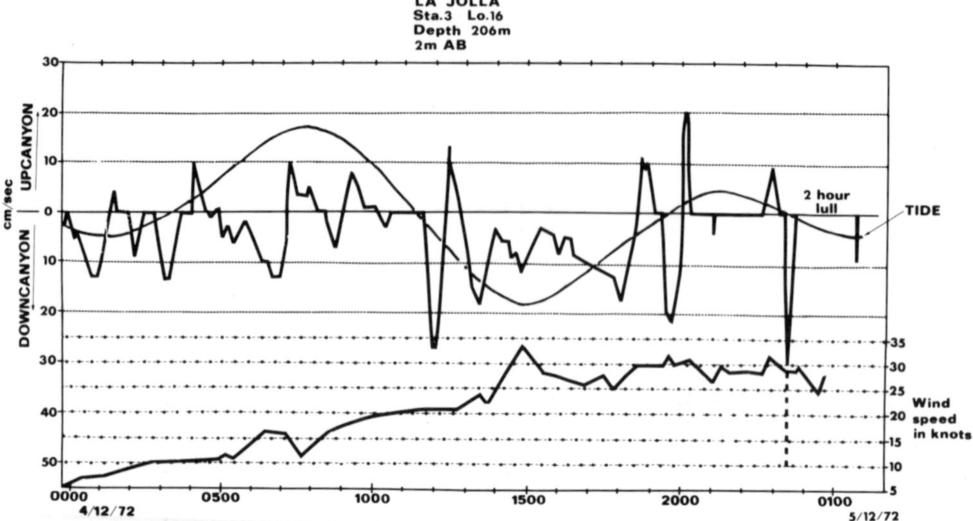

FIG. 20—Figure illustrates nature of currents preceding a turbidity current in La Jolla Canyon that carried away the current meters; wrapped in kelp, they were found by observers in *Nekton* submarine 0.5 km downcanyon from where they were originally dropped. The increase of wind velocity (obtained from U.S. Weather Bur.) is indicated. First small turbidity current is shown and apparently preceded the downcanyon transport (Fig. 21).

the erosion that apparently was produced by the turbidity current was an aluminum can of a recently marketed style. We learned that it was not more than 2 or 3 years old and was presumably much younger. This demonstrates the relative mobility of sediments versus time within canyons at depths of 150 to 180 m. This feature may be an example of the "chaotic facies" observed by Normark and Piper (1969) in the Miocene fanvalley formation known as the Doheny Channel. They referred to this facies as having developed by the alternating conditions of powerful erosion and active deposition.

Beginning in 1975, we worked in four areas off river mouths where considerable quantities of sediment are introduced into marine valley heads. In three of these areas we acquired records that strongly suggest low-speed turbidity currents. These currents all had maximum speeds on the order of 72 to 100 cm/sec (Figs. 22-27). The frequency of these low-speed turbidity currents, where rivers are introducing sediments near a canyon head, must be high. Aside from the single record off the Congo where turbidity currents are shown by breaking of cables, the only case of current measurements off a large river (where we did not record a turbidity current) was off the Fraser River of British Columbia; and there, constant dredging at the river mouth probably prevents frequent turbidity currents in the adjacent seavalley. Thus we might remove the Fraser Seavalley from this consideration owing to the acts of man. Although we made no attempt to choose periods when sediment introduction was especially large, the Rio de la Plata of Puerto Rico was in flood at the time of our observations (Fig. 97), and the operation off the Abra River of Luzon was during the rainy season with onshore winds. Thus there was probably more than usual chance for turbidity currents.

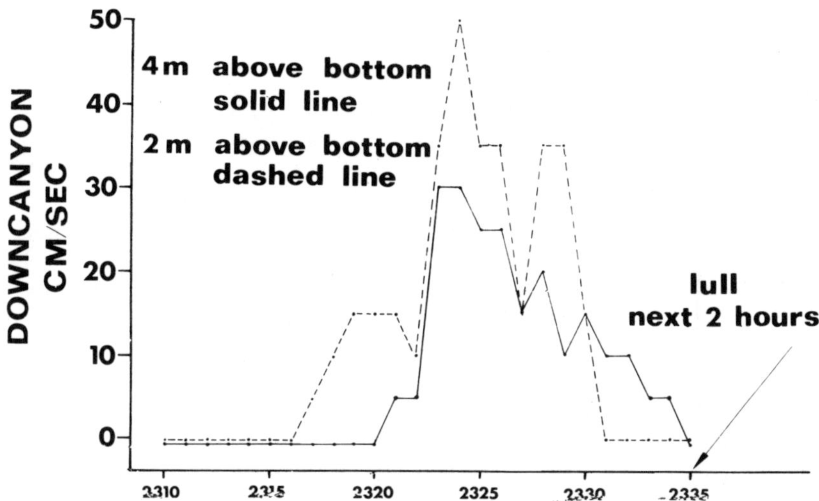

FIG. 21—Velocities of downcanyon current averaged for each minute during what may be only the early pulse of a turbidity current in La Jolla Canyon (Station 3; lowering 16; depth, 206 m). Note the rapid buildup to the peak velocity and slower decay followed by a period of no recorded current (Fig. 20).

Because we had our current meters emplaced for only a few days in each locality, it seems highly probable that turbidity currents occur many times each year at each of these locations. Off the Rio Balsas delta on the Mexican Coast, one might expect turbidity currents to be common during high swell periods, because during the only two periods when observations were made, these currents coincided with large waves (Reimnitz, 1971; Shepard et al, 1975).

Off large rivers where deltas are being built forward onto the seafloor slope, currents with velocities approaching 75 cm/sec probably are sufficiently powerful to cut valleys into the unconsolidated sediments and the foreset slopes.

Wherever the slopes off these advancing deltas were sounded, valleys or canyons of the kind shown in Figures 77 and 121 were found. If the same conditions existed in the past, then many unconformities in ancient deltaic sediments formed on prograding deltaic slopes. The slope does not have to be even as much as 1° (note the gullies off the Mississippi delta where slopes do not exceed 0.5° [Shepard, 1955]). Coleman and Garrison (1977) showed that sediments being introduced are unstable and cause slides with serious disruption to the numerous oil rigs in this area. These slides are known to start turbidity currents (James Coleman, personal commun.).

An examination of the one-minute integrated velocities (instead of the usual 5 or 7.5-minute average) of the turbidity currents gives some idea of their order of development (Figs. 21, 23, 25, 27). The two most complete records (Figs. 23, 25) differ in that one has only one strong surge (Fig. 25) and the other has a single surge followed

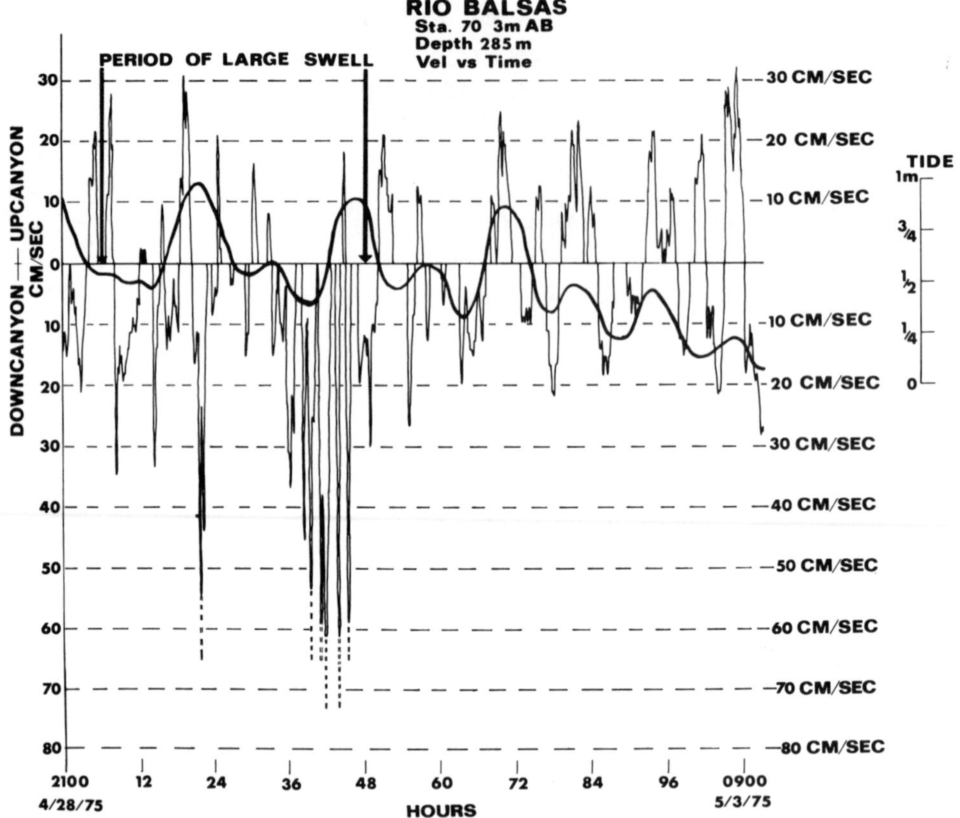

FIG. 22—Time-velocity record at 285 m in Rio Balsas Canyon. Strong downcanyon pulses indicate turbidity currents during the period of large swell and show a relation to high tide. For more detail see Figure 23. The relation of current alternations to tides in the last half of the record is surprising, for deeper stations (71) at 384 and (72) at 658 m do not show much tidal relation (Fig. 7).

after an interval of 12 hours by a group of about 6 surges (Figs. 22, 23). In all but Rio de la Plata Canyon (Fig. 27), these surges built up rapidly and declined rather slowly. In all four canyons prior to the turbidity current, there was a relatively strong upcanyon flow. This is shown particularly well in the Abra Seavalley and Rio de la Plata Canyon records (Figs. 24, 26). The time period after the turbidity-current surge when there is a lull is well shown in the Abra Seavalley record and after most of the Rio Balsas Canyon surges. The lull lasted only 15 minutes after one of the Rio Balsas surges while the record in Rio de la Plata Canyon did not indicate this time period because the current meter was set to release and came to the surface too soon.

Thus we appear to have a functional schematic of these slow turibidity currents. They apparently are related in most cases to an unusual buildup of water at the head of the canyon because an unusually high upcanyon current occurs just preceding each downcanyon turbidity flow. Like most unusual flows they generally develop rapidly to their fastest speed and then gradually decline. The natural result of a

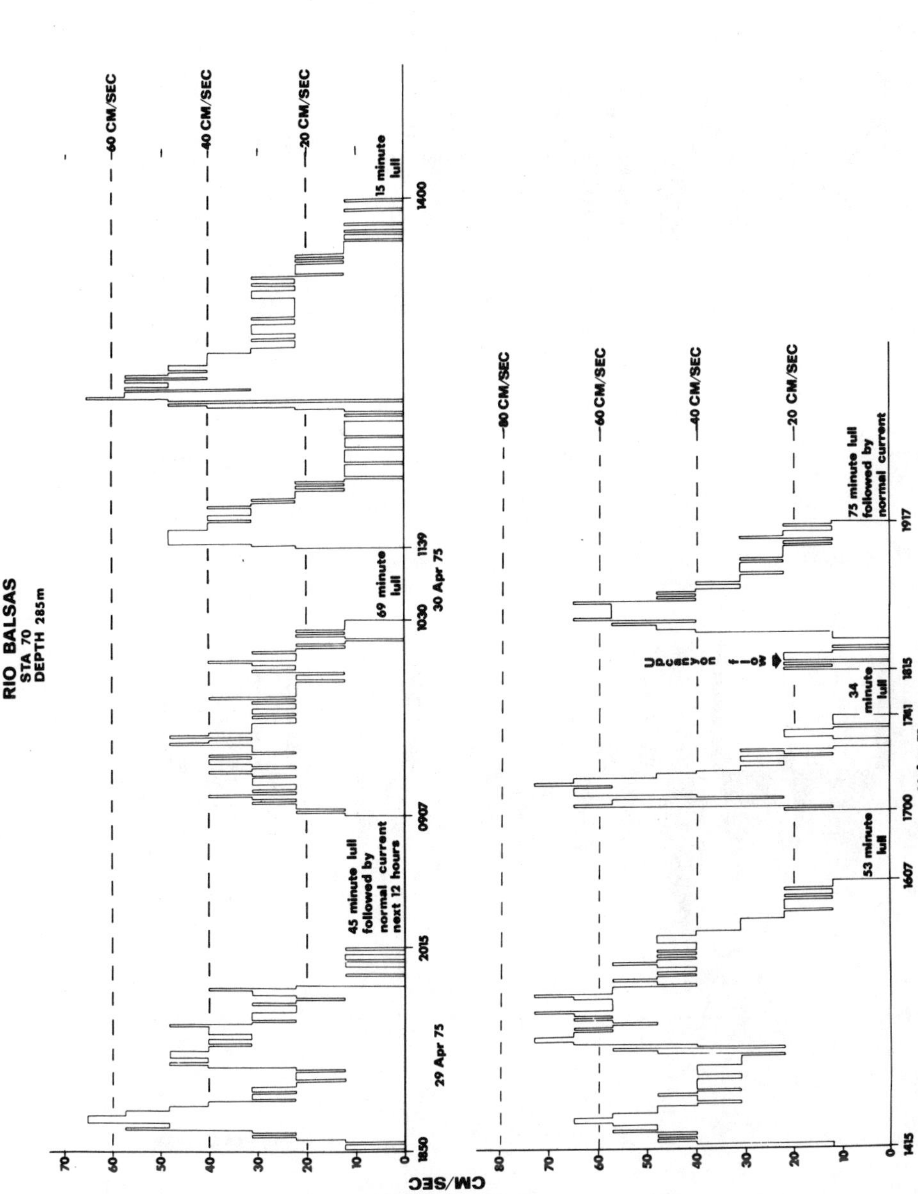

FIG. 23—One-minute current velocity averages during the strong downcanyon surges in Rio Balsas Canyon (Station 70; depth, 285 m). Note length of time with no measurable current following the flows. Also note that only one period of upcanyon flow was observed (at time 18:15). The steplike changes are due to nature of recording for 1-minute averaging, each step representing one additional mark on the tape during the 1-minute period (Fig. 22).

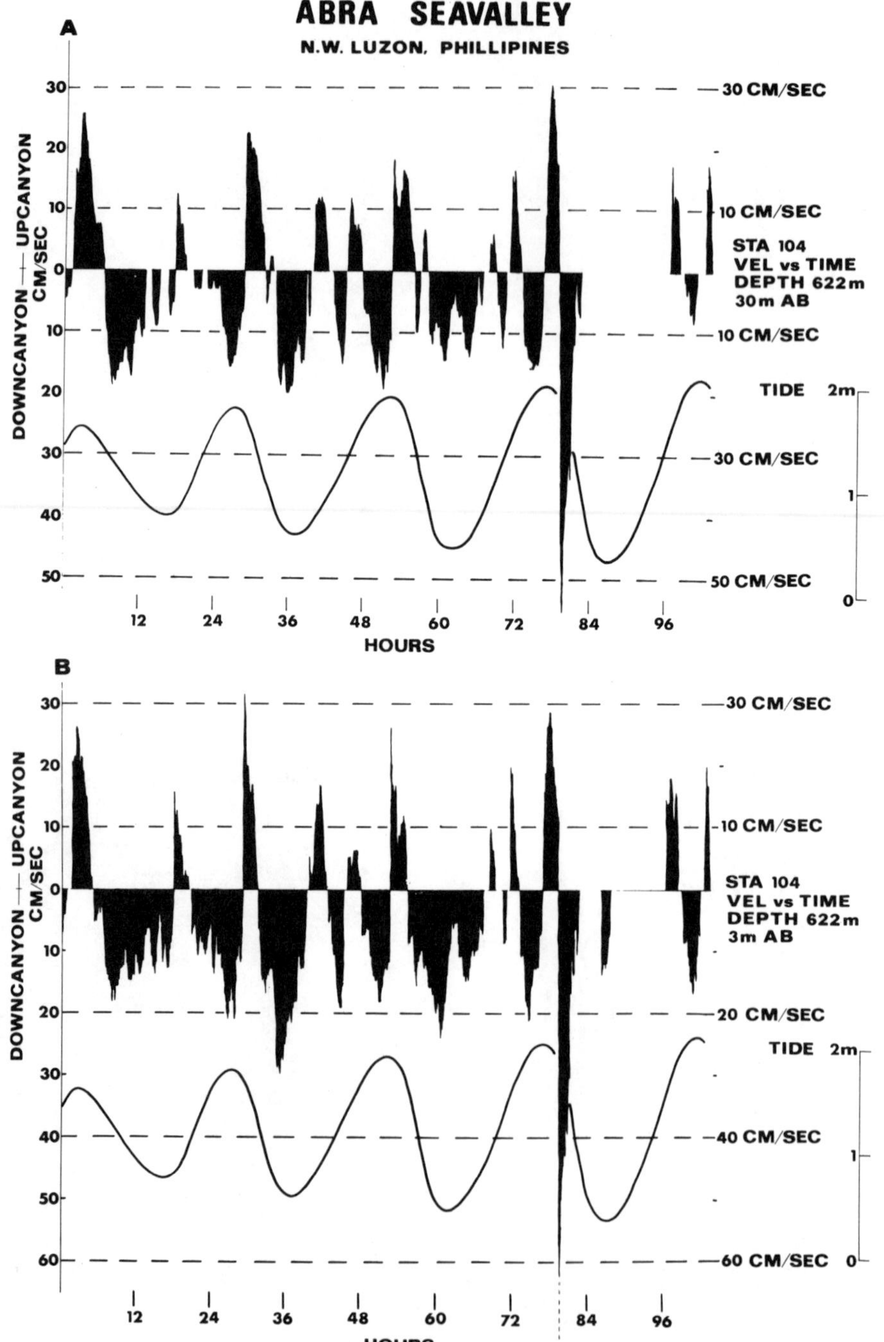

FIG. 24—Time-velocity curves from 3 and 30 m above bottom at 622 m in Abra Seavalley off the Abra River. Shows excellent example of a turbidity current at hour 79 at both levels. Note the upcanyon flow that just preceded the turbidity current and the long period without current that followed. Because the turbidity current is indicated at 30 m above bottom (upper section), the current had considerable thickness (Fig. 25).

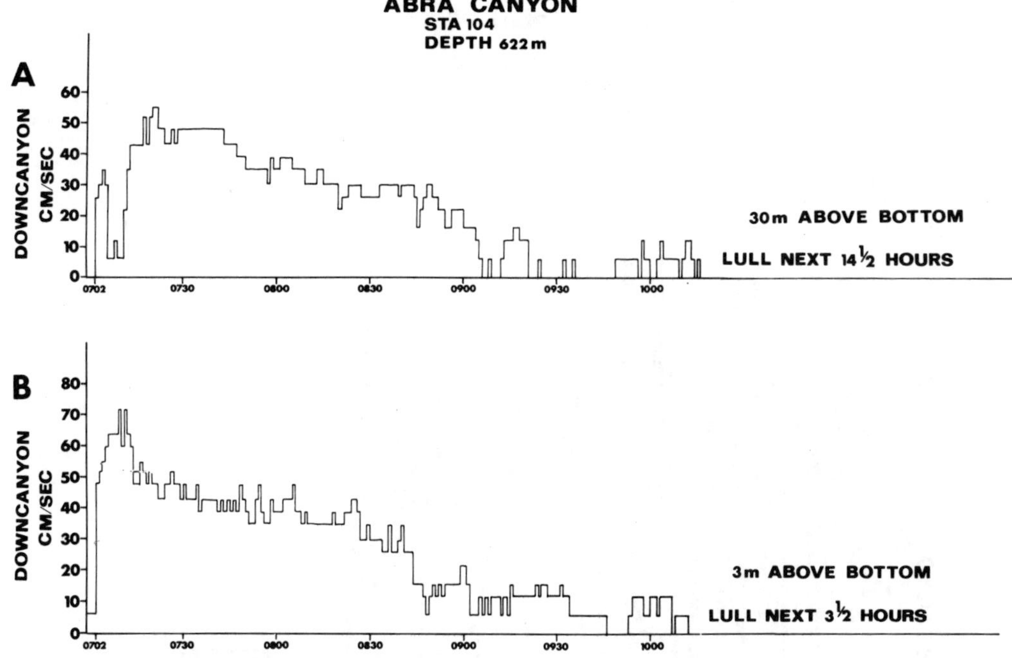

FIG. 25—One-minute averages for current velocities during the Abra Canyon turbidity current (Station 104; depth, 622 m). Note rapid buildup up the peak and the slow decay to the period of no measurable velocity. **A**, 30 m above the bottom; downcanyon flow followed by amazingly long period of no velocity (lull = 14.5 hours). **B**, 3 m above the bottom; downcanyon flow followed by 3.5-hour lull (Fig. 24).

turbidity current is to disturb the usual tidal and internal wave sequence, so that the lull following the turbidity flow is not difficult to understand.

Our data provide a few indications as to the distances to which these turbidity currents flow. Apparently the current at Rio de la Plata off northern Puerto Rico does not continue for the 2 km to the deeper station (Fig. 96). And the currents off Rio Balsas do not appear to have moved down to a deeper station in the canyon at a distance of 20 km. On the other hand, when we permanently lost a current meter in Scripps Canyon at 138 m, another also disappeared approximately 1.2 km farther down the canyon at 260 m, suggesting that this unmeasured current flowed practically the length of Scripps Canyon. The finding of our two current meters 0.5 km downcanyon in La Jolla Canyon at the time of the 1972 turbidity current suggests that the flow lost much of its transporting power in half a kilometer. Thickness of the turbidity currents is shown only by the fact that we had current meters at both 3 and 30 m in Abra Canyon in northwest Luzon (Fig. 24) and the flow was almost as powerful at 30 m as at 3 m above the bottom. The finding of coarse sediment on the levees of fanvalleys several hundreds of meters above the bottoms of the valleys (Shepard, 1966) further suggests that turbidity currents have a considerable thickness, but obviously more information is needed.

The discovery that these turbidity currents may occur with considerable frequency (at least in areas where large amounts of sediment are being introduced) certainly

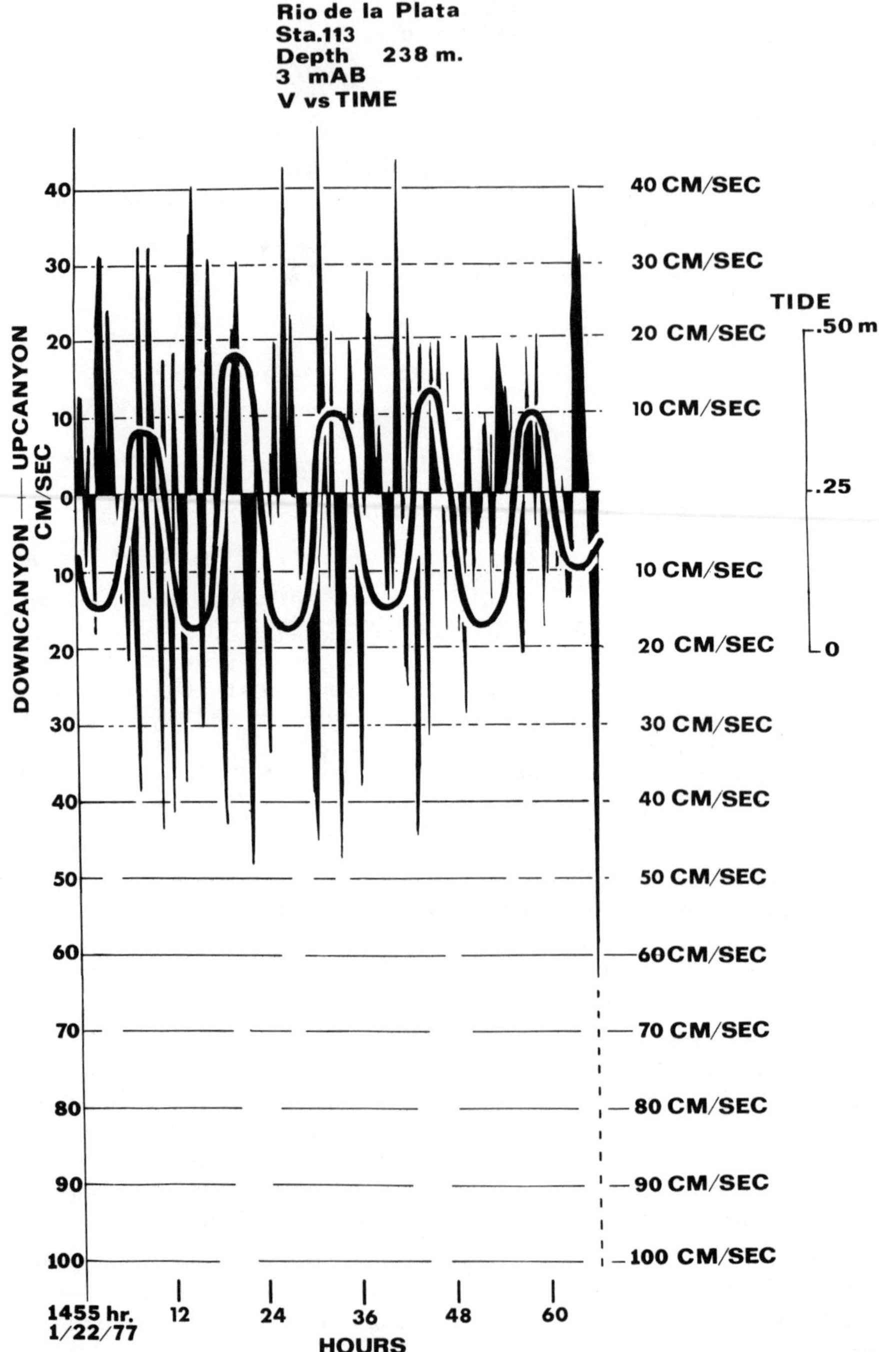

FIG. 26—Time-velocity curve from Rio de la Plata Canyon at 238-m depth. Unusually fast currents with short alternation cycle period. Currents have fastest average of up- and down-canyon flows of any record to date. Apparently the record terminated during a turbidity current (Fig. 27).

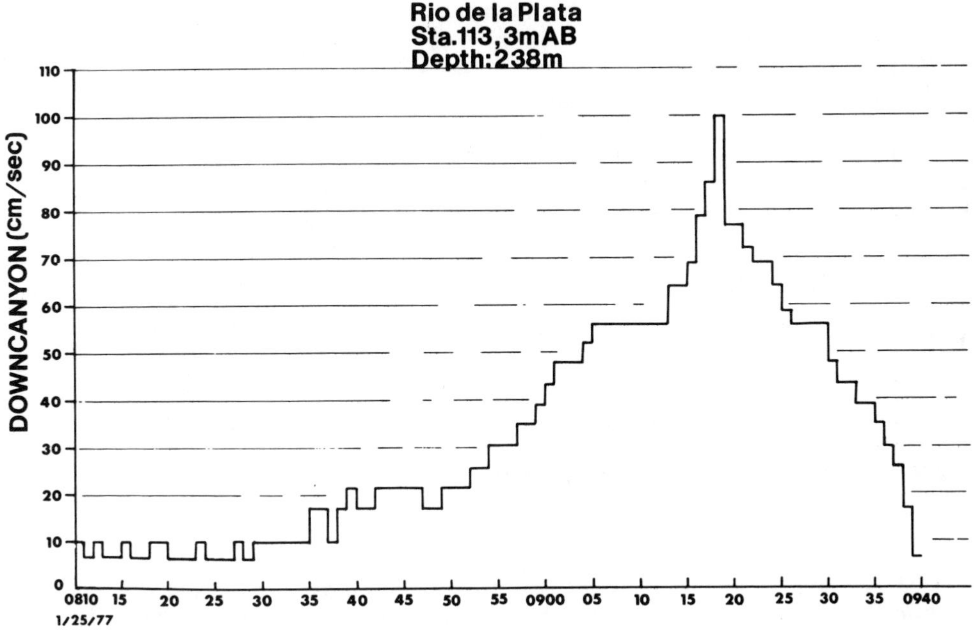

FIG. 27—One-minute velocity averages for the strong surge in Rio de la Plata Canyon. This example of a turbidity current was unlike others in showing somewhat faster decay than buildup of the maximum speed. The record terminated immediately after the turbidity current leaving no time for a lull to be recorded.

indicates the importance of turbidity currents in general. This was suggested by Heezen et al (1964) relative to the Congo. Some earlier work (Dill, 1964; Shepard et al, 1969) indicated that turbidity currents are relatively rare. This may be true of the very fast currents caused by large slides or earthquakes such as that off the Grand Banks in 1929 (Heezen and Ewing, 1952), but more recently Moore (1969) observed turbid sea water moving slowly down submarine canyons from a deep-diving vehicle off the La Jolla area. He explained these seaward drifts as very slow but almost continuous turbidity currents. Marshall (unpub.) observed stratified layers of turbid water in Scripps Canyon while deep diving in the *Nekton Gamma.* Drake and Gorsline (1973) showed that bodies of turbid water move up- and downcanyon with the normal currents. Somewhat faster currents (30 cm/sec) are certainly due to density flows in the lakes of Switzerland. They correspond to the pulses of river flow coming into lakes such as the Walensee (Fig. 28; Lambert et al, 1976). Also the observations of Inman et al (1976) showed that edge waves cause pileup of water along the shore at the head of canyons and may cause relatively weak currents to move down the canyon floors and (occasionally) cause currents of greater strength.

The recent work on slumps along the flanks of Scripps Canyon indicates that they probably occurred rather frequently, possibly owing to differential wave pressures during storms (Marshall, 1978). These slumps that provide up to 100,000 cu m of sediment as lateral infill from the adjacent shelf probably turn into turbidity flows transporting sediments down the canyon a distance proportional to their energy input plus the impetus of storm or high-swell-induced flows within the canyon.

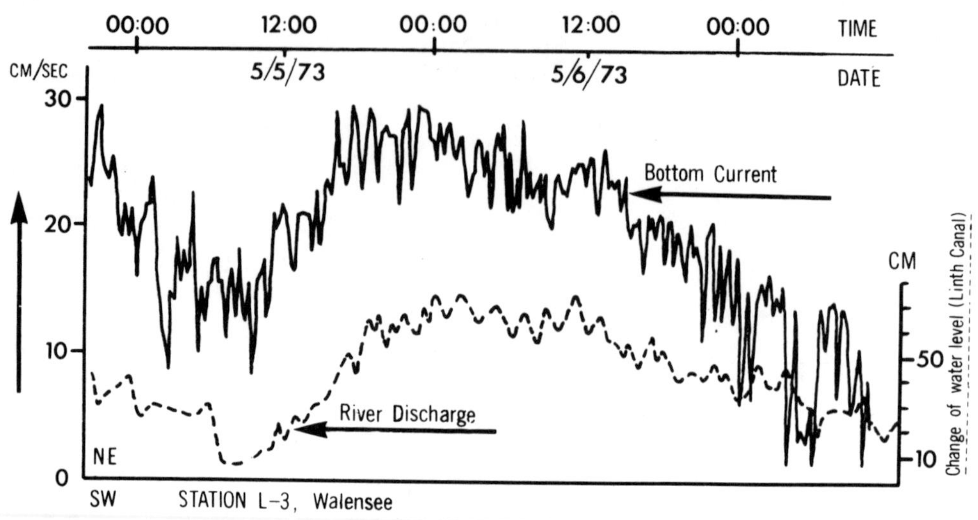

FIG. 28—Pulsating downslope flow in Lake Walensee, Switzerland, compared with neighboring river flow. Record obtained by Neil Marshall (Lambert et al, 1976).

The new emphasis (Coleman and Garrison, 1977) on the discovery of slumps occurring along the Mississippi delta region adds further credence to the theory that turbid flows occur more frequently, at least in some environments, than was previously supposed.

With the exception of man-made turbidity currents (Normark and Dickson, 1976a, b), the only other example we have of turbidity currents actually measured by current meters comes from the Var Canyon off southern France (Fig. 12) where the currents measured for 1 minute every 5 minutes appeared to reach velocities up to 92 cm/sec (Gennesseaux et al, 1971), but currents flowed continuously downcanyon, as mentioned previously.

Fraser Delta Seavalley and Congo Canyon

The comparison between sedimentation off the mouths of two large rivers, Congo of Africa and Fraser of western Canada, is pertinent in this discussion of marine valleys. They both introduce large masses of sediment into the ocean, but the Fraser is building its delta forward rapidly (Mathews and Shepard, 1962) whereas the Congo has an estuary coming well up into its mouth and a submarine canyon that extends seaward (Fig. 123). The Congo frequently has large turbidity currents that break cables laid across the canyon (Heezen et al, 1964). Our current-meter measurements failed to detect any clear evidence of turbidity currents (Fig. 124), but this was probably because of the relatively short period of observation. In any case the Congo mouth is exceptional, as there is no other large river where turbidity currents and longshore currents are entirely successful in transferring all of the sediment from the river down a canyon to the deep underwater fan that is located at this position (Fig. 123). The Fraser mouth is far less effective in producing turbidity currents from the piling up of sediment where it meets the sea, and only a very shallow seavalley (Fig. 71) exists at this place. The dredging at the Fraser mouth may, as mentioned previously, inhibit turbidity-current development. No doubt the difference in sediment type is also part of the explanation of the contrast.

Currents and Origin of Submarine Canyons

It would be presumptuous to claim that a study such as ours of canyon currents provides the answer to the problem of why the continental slopes of the world are incised almost everywhere by deep canyons; many of them cut into hard rock and most of them extend to great depths. Of course, Daly wrote more than 40 years ago (1936) that the canyons were cut by turbidity currents, but Daly really did not have any actual observations on which to base his ingenious hypothesis, and in fact the many scientists who adopted his idea at a later period were not using much more than speculation on which to base their theories. Our studies of the currents in the canyons at least provide a somewhat firmer foundation, and we hope this will prove useful.

The most important concept that our investigation of currents has shown, relative to canyons, is that canyonlike valleys can be formed by turbidity currents on the fronts of deltas which are building out onto the continental slope. Here we have evidence of a very favorable environment for slumping and for turbidity currents. Furthermore, we find not only that small valleys like those off passes of the Mississippi delta (Shepard, 1955), but also features which apparently are developing into true canyon proportions like those off the Rio Balsas delta and the Abra delta (Figs. 77, 121), were probably developed in this manner.

Where sea levels were lowered during the glacial stages of the Pleistocene, most rivers were said to be carrying larger quantities of sediment than they do now because of the generally higher precipitation and because of steeper slope gradients that most likely existed at a former shelf edge. Rivers were able to build deltas along the shelf margins, and many canyons began at this time. Many of these valleys would have remained open after the rise in sea level, because it is clear now that currents still exist even in submarine canyons far from any river source of sediment (like the New England canyons off Georges Bank). Strong currents in the New England canyons were measured, and a turbidity current was observed from a deep-diving vehicle by Richard Slater (personal commun., 1975) in Oceanographer Canyon.

It is unwise to think that these promising clues to the origin of canyons are the final solution of all the problems which the canyons present. Surely some canyons were formed originally as river valleys and later submerged to where marine processes took over their subsequent development. It is for areas where there is no apparent indication of great submergence, and where submarine canyons are common, that we need to develop hypotheses that include the development of valleys on the front of advancing deltas.

Valleys have various other causes, including faulting and folding. Large landslides produce many valleys on the lands and may account for the initiation of some of the submarine canyons and even of many tributaries (see, for example, Marshall, 1978). Ordinary currents are strong enough so that they may keep the valleys from filling in many localities, and turbidity currents are frequent in these valleys wherever they have abundant sources of sediment. Although out studies have only shown relatively weak turbidity currents, these currents are at least of sufficient strength to transport large quantities of sediment down the canyons. Furthermore, indirect evidence (the breaking of cables laid across canyons) suggests occasional currents of much greater strength, especially during large earthquakes. These more powerful currents still offer a challenge and should be investigated further. To date our attempts have not brought success, but we have not exhausted the fields of investigation.

Crossvalley Currents

It is easy to understand how currents normally flow in a direction along a valley axis, and features on the valley floor and walls can commonly cause a flow that is decidedly altered from the canyon axial trend. However, where we find that crossvalley flows are a major feature, as during one period in Hueneme Canyon at two stations (Figs. 47-49) and at almost all stations in Monterey Canyon (Fig. 62), we are considerably mystified.

In Hueneme Canyon, it is possible that the strong crosscanyon flows were related to the strong northwesterly winds prevalent during the data-collection period, and to the consequent movement (past the area of Hueneme Canyon) of large masses of water out of Santa Barbara Channel (Fig. 46). It is not easy to see how this water mass can affect the flow along this relatively deep canyon floor, but it is possible that vertical eddies may be formed in a crosscanyon direction owing to the shear of the upper water mass moving across the canyon normal to its axis. Perhaps by coincidence, the strongest crosscanyon flows occur during periods of low tide (Fig. 48). The flows in Monterey Canyon have little if any relation to canyon axial directions, which is quite contrary to those in adjacent Carmel Canyon; this leaves us completely puzzled, as we are with many other characteristics of Monterey Canyon currents. At Monterey Canyon there is little to indicate that crosscanyon flows are related to tides or to wind directions. Here a large-scale investigation seems to be required before answers are likely to become available.

In Hudson Canyon, currents at our deep station showed an almost complete disregard of the canyon axial trend (Fig. 89B). Where a valley has a very broad floor, as in the Kaulakahi Channel of northwest Kauai, it is not surprising that the currents show little relation to the axis, but Hudson Canyon (according to profiles) is relatively narrow at this location. Perhaps in canyons with wide floors, canyon currents meander along the canyon axis as a small stream meanders in a wide valley. No other explanation is evident to us unless, as suggested earlier, cross-axis eddies can be formed by regional water-mass movements.

In some cases the polar plots showed an apparent flow axis at a different canyon direction from that indicated by charts (Figs. 80, 84). We think that in most of these cases the available charts may fail to show smaller scale axial direction variations properly. More information is desirable for a better understanding of this phenomenon.

Internal Waves Advancing Along Valley Axes

It was discovered that the time-velocity pattern of up- and downcanyon flows at two adjacent stations in Carmel Canyon showed close resemblance to each other after one was shifted timewise until it fitted the other (Fig. 60A). This was our first definite evidence that internal waves are moving along the canyon axes and correspond to the movement of internal waves across the continental shelf (LaFond, 1962). Further evidence came from an analysis by Gordon and Marshall (1976). We now find many cases where we can match these patterns (although only roughly in some instances). Because internal waves are known to advance toward the shore across continental shelves, we expected internal waves to be moving up the axis of submarine valleys. However, when we worked in Santa Cruz Canyon and looked for a match for the curves, we found that the best match of the two adjacent stations was made when the patterns matched using a later time for the downcanyon station, suggesting that the internal wave was moving in that direction (Fig. 54A). This was confirmed in Santa Cruz Canyon by finding that the same thing happened on a second occasion (Fig. 54B).

Similar results are obtained from other data where the phase lead (or lag) of one series over the other is calculated for the most predominant of the component sine

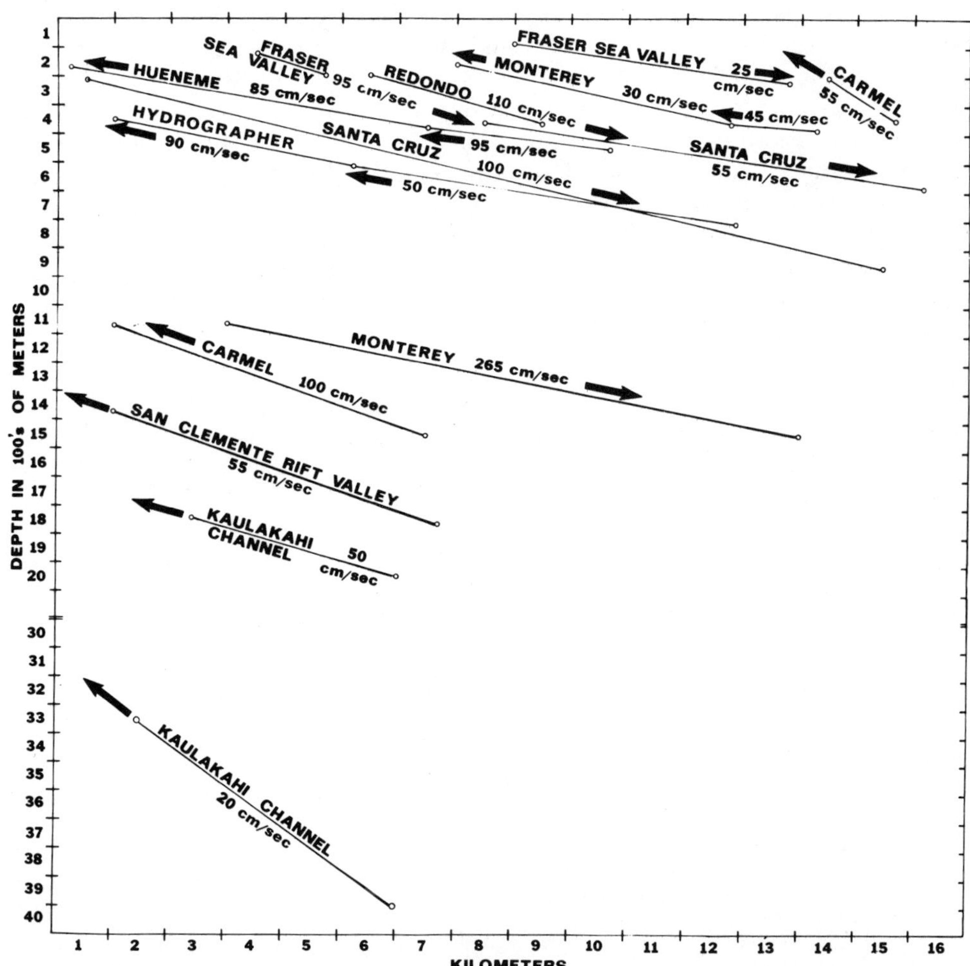

FIG. 29—Profiles showing the probable direction of advance of internal waves along the canyon axes in various canyons where contemporaneous records were made at two or more axial depths. Rate of advance could only be determined roughly and even the direction of advance is not entirely certain in some cases. Vertical scale × 5.

waves that compose the time series. The phase, together with distance between stations, can give an estimate of wave speed and direction.

From our data we find that we have 20 examples where internal waves appear to advance up canyons or other types of marine valleys, and seven cases where they appear to advance downcanyon (Fig. 29). In Monterey Canyon where most of our results are contrary to the usual findings, we discovered two cases of adjacent stations where the advance was upcanyon and one where the advance was downcanyon, although in the latter the advance was at a different time. In other cases where we have more than one pair of stations from the same canyon, advance was always in the same direction. Thus we have three pairs each for comparison in La Jolla and Hueneme Canyons and also in Kaulakahi Channel and two pairs in Hydrographer Canyon, all advancing upcanyon, and two pairs in Fraser Seavalley advancing downcanyon.

It certainly is significant that in all but two of the cases where apparent downcanyon advance of internal waves occurred there is clear evidence of the introduction of water-mass flow into the head of the canyon or seavalley that might have caused internal waves to progress in that direction. Thus the head of Santa Cruz Canyon is in the strait between Santa Cruz and Santa Rosa Islands where large masses of water are funneled southward from Santa Barbara Channel into the Santa Cruz Basin by northwest winds (Fig. 46). Downvalley internal-wave advance also is well established off the Fraser River, which is the largest river of western North America and has a considerable volume of fresh water entering at the seavalley head. Similarly, the phase lag off Rio Balsas, the largest west-coast Mexican river, seems to indicate downcanyon advance.

Appendix: Table 1. Summary of Current-Meter Data.

Canyon	st-lo-ser	Depth (m)	Fig. No.	Height Above Bottom (m)	Average Velocity (cm/sec)	Max. Vel. (cm/sec) Up	Max. Vel. (cm/sec) Dwn	Average Velocity (cm/sec) Up	Average Velocity (cm/sec) Dwn	Average Velocity (cm/sec) Cross	Average Duration (hrs) Up	Average Duration (hrs) Dwn	Average Duration (hrs) Cross	Direction of Net Flow	T*	Avg. Cycle Length (hrs)	Tidal Range (ft)	Tidal Range (m)
La Jolla	01-01-01	46	13	3	3.8	7	10	4.0	3.7	3.8	1.3	0.8	0.4	up	42	2.1	5.0	1.6
	01-02-64	46	13	3	5.7	32	16	7.2	5.7	4.3	2.6	1.0	0.4	up	187	3.8	9.6	3.0
	02-01-02	107	13	3	5.4	23	10	4.4	4.0	7.8	0.9	0.6	0.8	up	58	3.7	8.0	2.5
	03-01-03	206	13	3	7.9	18	23	8.0	10.9	4.8	1.3	2.2	0.3	down	63	4.1	5.9	1.8
	03-03-07	206	8, 11, 13	3	8.5	22	27	9.4	9.4	6.6	1.5	1.8	0.4	down	123	3.7	6.5	2.0
	03-05-11	206	13	3	6.5	18	24	6.1	9.4	3.9	1.5	1.8	0.3	down	66	4.0	9.1	2.8
	03-06-14	206	13	3	4.1	15	19	4.8	5.6	1.9	1.1	1.9	0.3	down	110	4.0	7.2	2.2
	03-07-17	206	13	3	3.5	13	26	4.6	3.7	2.2	1.3	1.9	0.3	down	99	3.3	7.0	2.1
	03-08-20	206	13	3	3.8	13	21	4.4	4.9	2.2	1.1	1.5	0.4	down	123	3.4	5.1	1.6
	03-09-23	206	13	3	7.8	16	23	6.0	11.0	6.3	1.1	1.3	0.9	down	94	3.5	6.8	2.1
	03-10-27	206	13	3	8.0	23	32	8.1	12.5	3.4	1.2	1.7	0.3	down	98	3.3	6.2	1.9
	03-11-29	206	13	3	4.4	13	18	3.5	5.9	3.7	0.7	1.4	0.9	down	114	3.5	6.5	2.0
	03-12-35	206	13	3	4.2	15	27	4.0	6.4	2.2	1.1	1.9	0.4	down	93	3.7	8.3	2.6
	03-13-40A	206	9, 10, 13, 14	3.6	6.5	17	26	6.0	11.5	2.9	1.1	1.5	0.5	down	103	3.7	6.2	1.9
	03-13-40B	206	9, 13	19	8.9	19	24	9.1	13.9	3.7	1.4	1.4	0.8	up	103	6.0	6.2	1.9
	03-13-40C	206	9, 10, 13	34	4.6	12	24	4.8	7.2	1.9	1.8	2.2	0.6	down	103	7.1	6.2	1.9
	03-14-42	206	13	3	4.8	16	26	5.2	6.6	2.6	1.1	1.1	0.3	down	104	3.3	8.6	2.7
	03-15-73A	206	9, 13,15	3.6	4.4	13	21	4.9	6.6	1.8	1.0	1.2	0.5	down	214	3.6	6.6	2.0
	03-15-73C	206	9, 13, 15	11	4.3	14	21	4.7	6.6	1.6	1.2	1.2	0.5	down	214	3.7	6.6	2.0
	03-15-73D	206	9, 13, 15	34	4.1	18	25	4.5	6.1	1.6	1.1	1.1	0.6	down	214	4.5	6.6	2.0
	03-16-74A	206	13, 16, 121	2	4.1	21	30	4.2	7.4	0.8	0.5	1.2	0.2	down	84	2.4	7.2	2.2
	03-16-74B	206	13, 121	4	3.1	26	24	3.3	5.2	2.4	0.7	0.3	0.1	down	84	2.7	7.2	2.2
	03-17-97	206	5, 13	3	5.2	18	28	4.9	8.8	2.0	1.3	1.7	0.2	down	88	3.6	8.2	2.5
	03-18-133B	206	13	30	3.5	18	10	5.2	3.7	1.6	2.3	2.2	0.7	up	92	6.6	7.6	2.4
	03-19-194A	206	13	3	4.7	20	29	5.0	8.0	1.2	0.8	0.8	0.1	down	228	9.0	8.8	2.7
	03-19-194B	206	13	50	1.5	14	18	2.0	2.0	0.5	1.1	0.9	0.2	down	228	8.7	8.8	2.7
	03-19-194C	206	13	100	4.5	17	16	5.9	6.0	1.7	1.2	1.0	0.2	up	218	12.0	8.8	2.7
	05-01-06	167	11, 13	3	5.1	21	21	6.2	5.3	3.8	0.7	1.3	0.4	down	120	4.0	6.5	2.0
	05-02-10	167	13	3	**	**	**	**	**	**	1.0	1.4	0.3	**	135	3.1	8.5	2.6
	05-03-13	167	13	3	**	**	**	**	**	**	0.9	1.0	0.4	**	140	2.3	6.6	2.0

*Total duration in hours. **Direction data only; no velocity data. ***Velocity data only; no direction data.

Canyon	st-lo-ser	Depth (m)	Fig. No.	Height Above Bottom (m)	Average Velocity (cm/sec)	Max. Vel. (cm/sec)		Average Velocity (cm/sec)			Average Duration (hrs)			Direction of Net Flow	T*	Avg. Cycle Length (hrs)	Tidal Range	
						Up	Dwn	Up	Dwn	Cross	Up	Dwn	Cross				(ft)	(m)
	05-04-16	167	13	3	3.1	15	15	4.2	3.7	1.4	1.1	1.4	0.4	down	102	5.0	7.2	2.2
	05-05-18	167	13	3	6.6	21	21	6.6	8.5	4.8	1.2	1.3	0.4	down	94	3.6	7.0	2.1
	05-06-21	167	13	3	5.1	21	24	6.0	5.9	3.3	1.1	1.4	0.3	down	97	3.4	5.1	1.6
	05-07-24	167	13	3	5.4	21	24	5.5	7.2	3.7	0.8	1.4	0.3	down	95	3.2	6.8	2.1
	05-09-31	167	13	3	6.3	22	9	7.2	5.9	5.9	0.6	1.5	1.0	down	104	5.5	6.5	2.0
	05-11-43	167	13	3	6.1	24	24	7.2	6.6	4.5	0.9	1.2	0.3	down	103	3.2	8.6	2.7
	05-12-134	167	13	3	5.3	15	21	5.6	6.8	3.4	1.1	1.2	0.7	down	103	3.2	8.6	2.7
	06-01-32	78	13	3	3.9	15	18	4.3	4.2	3.1	1.1	0.8	0.4	up	126	3.6	6.5	2.0
	06-02-38	78	13	3	2.1	7	10	2.6	2.5	1.2	1.1	0.7	0.4	up	82	4.1	8.3	2.6
	06-03-95	78	13	3	3.6	17	12	3.9	4.1	2.7	1.2	0.8	0.4	up	87	2.8	8.2	2.5
	10-01-45	375	13	3	3.0	10	13	2.9	3.6	2.4	2.4	3.0	0.4	down	108	9.5	8.5	2.6
	10-02-46	375	13	3	9.2	24	21	10.1	11.0	6.5	3.5	3.8	0.6	down	105	10.6	7.5	2.3
	10-03-49	375	13	3	4.3	15	13	4.6	4.5	3.8	1.7	2.6	0.8	up	109	9.5	5.5	1.7
	10-04-98	375	5-7, 13	3	9.3	27	25	12.0	11.0	5.0	4.6	4.4	0.8	down	84	11.3	8.2	2.5
	44-01-96	135	13	3	4.1	17	24	4.8	6.0	1.6	0.8	1.0	0.3	down	88	4.4	8.2	2.5
Scripps	09-01-41	125	13	3	2.8	12	9	2.6	3.3	2.5	1.0	1.1	0.4	down	76	4.1	8.6	2.7
	09-02-44	125	13	3	5.4	24	32	5.8	6.0	4.4	0.3	0.7	0.3	down	114	2.4	8.5	2.6
	100-01-162A	40	13	3	4.2	12	10	4.7	4.6	3.3	0.6	0.8	0.4	down	335	2.9	7.9	2.4
	100-01-162B	40	13	10	3.1	7	7	3.2	3.2	3.0	1.2	0.6	0.4	up	335	3.0	7.9	2.4
	107-01-169A	39	13	3	3.1	8	9	3.6	3.2	2.6	0.6	0.6	0.3	up	283	1.6	6.0	1.9
	107-01-169B	39	13	13	4.8	7	8	5.1	4.8	4.5	0.6	0.5	0.2	up	283	2.0	6.0	1.9
	108-01-170A	39	13	3	4.0	6	8	4.3	4.2	3.4	0.5	0.3	0.6	up	238	4.5	6.0	1.9
	108-01-170B	39	13	13	5.9	20	17	6.4	7.0	5.5	1.0	0.5	0.4	up	238	2.7	6.0	1.9
Newport	20-01-60	101	17	3	4.6	13	18	6.2	4.8	3.0	1.2	1.5	0.4	down	46	4.4	6.0	1.9
	20-02-62	101	17	3	**	**	**	**	**	**	1.5	1.9	0.4	**	68	4.0	6.0	1.9
	21-01-61	252	17	3	3.5	13	10	4.5	3.3	2.7	2.1	2.9	0.3	up	46	11.0	6.0	1.9
	61-01-115A	366	17, 18	3	6.0	26	21	7.5	8.5	2.1	3.1	5.5	0.6	down	309	10.9	8.1	2.5
	61-02-145A	366	17	3	3.0	9	11	4.2	3.7	1.1	4.0	2.9	1.6	up	410	10.8	2.2	0.7
Santa Monica	23-01-68	458	23	3	9.4	27	30	12.8	10.5	4.8	4.8	8.5	1.9	down	67	16.0	8.8	2.7
Redondo	18-01-58	95	19, 20	3	4.9	24	24	6.2	6.4	2.2	1.6	1.9	0.4	down	369	5.7	5.0	1.6

	18-02-65	95	19	3	3.9	13	10	5.2	4.4	2.2	1.8	1.8	0.5	down	69	5.0	4.8	1.5
	19-01-59	274	19	3	**	**	**	**	**	**	2.0	2.6	0.9	**	520	5.0	5.0	1.6
	19-02-66	274	19	3	3.1	21	10	4.6	3.3	1.3	3.6	1.1	0.4	up	69	12.4	4.8	1.5
	19-03-86A	274	19	3	6.4	21	18	6.6	6.8	5.7	3.4	3.9	1.0	down	90	10.8	4.2	1.3
	19-03-86B	274	19	90	6.1	20	20	6.6	8.3	3.3	3.5	3.7	1.2	down	90	10.8	4.2	1.3
	36-01-85	190	19, 21, 22	3	5.6	33	10	8.9	4.4	3.6	1.0	1.2	1.4	up	91	5.6	4.2	1.3
	37-01-87A	362	19, 22	3	6.5	30	27	8.1	8.0	3.3	2.0	2.0	0.9	down	89	5.7	4.2	1.3
	37-01-87B	362	19, 22	33	11.0	17	11	12.5	15.6	4.8	2.7	1.7	1.3	up	89	5.7	4.2	1.3
Hueneme	26-01-71	174	24	3	**	**	**	**	**	**	1.3	2.2	0.3	**	84	5.5	6.7	2.1
	26-02-75	174	24, 27	3	7.8	26	23	10.9	9.7	2.8	1.9	2.0	0.5	up	82	5.1	7.7	2.4
	27-01-72	375	24	3	13.6	34	34	15.3	19.7	5.7	3.3	5.8	0.6	down	84	11.5	6.7	2.1
	27-02-76	375	24-27, 29	3	20.4	35	30	21.1	18.7	21.4	2.9	1.8	1.2	up	84	8.0	7.7	2.4
	28-01-77	448	24, 27-29	3	17.7	32	32	23.6	21.1	8.3	4.3	4.9	0.4	down	85	11.6	7.7	2.4
	78-01-135	362	24	3	9.7	30	32	9.4	14.7	4.9	2.5	5.3	1.2	down	521	10.0	7.7	2.4
	84-01-141	201	24	3	7.7	24	23	10.1	9.1	3.9	1.6	3.2	0.9	down	100	7.4	5.5	1.7
	85-01-142A	256	24, 30	3	10.5	24	34	14.2	12.8	4.7	2.6	3.2	1.2	down	200	5.9	5.5	1.7
	85-01-142B	256	24, 30	30	7.9	26	33	8.2	11.7	3.7	2.5	3.0	2.1	down	235	9.2	5.5	1.7
	86-01-143	406	24	3	11.5	31	33	13.9	15.6	5.1	3.1	5.3	1.2	down	234	11.5	5.5	1.7
Santa Cruz	29-01-78A	357	31-34	3	9.8	36	24	13.6	12.2	3.7	3.4	3.7	2.1	down	80	10.8	5.9	1.8
	29-01-78B	357	31, 32, 34	29	9.3	23	25	13.0	10.4	4.6	3.2	2.7	1.2	down	80	9.5	5.9	1.8
	30-01-79	585	31, 33	3	6.9	22	29	7.6	10.7	2.3	1.5	4.8	0.5	down	80	9.8	5.9	1.8
	81-01-138	208	31, 33	3	11.4	35	38	12.8	16.0	5.3	2.1	2.5	0.8	down	81	5.8	5.5	1.7
	83-01-140	860	31, 33	3	7.7	23	31	7.7	12.2	3.2	1.6	2.5	0.7	down	70	5.8	5.5	1.7
Carmel	11-01-47	156	35, 40	3	5.5	15	20	5.0	7.6	4.0	0.7	1.8	0.7	down	119	2.7	5.5	1.7
	31-01-81A	205	35, 39, 40, 119	3	10.4	28	32	12.4	14.5	4.3	0.6	1.3	0.3	down	71	4.2	5.2	1.6
	31-01-80B	205	35, 40, 119	19	9.1	27	30	10.7	12.9	3.8	0.9	1.3	0.3	down	71	4.1	5.2	1.6
	32-01-81A	348	36, 37, 39, 40	3	14.4	15	36	15.8	19.5	7.9	1.0	1.8	0.9	down	70	4.6	5.2	1.6
	32-01-81B	348	36-38, 40	30	11.4	31	33	15.4	14.0	4.9	1.1	1.7	0.5	down	70	5.7	5.2	1.6
	56-01-110A	1070	39, 40, 119	3	9.6	27	31	9.6	15.0	4.3	2.5	3.1	0.7	down	96	11.5	6.5	2.0
	56-01-110B	1070	40, 119	30	9.0	21	31	10.7	12.8	3.4	3.5	3.4	0.8	down	96	11.6	6.5	2.0
	57-01-111A	1445	39, 40, 119	3	8.9	23	23	11.3	10.3	5.2	3.6	5.5	1.1	down	98	11.6	6.5	2.0

*Total duration in hours. **Direction data only; no velocity data. ***Velocity data only; no direction data.

Canyon	st-lo-ser	Depth (m)	Fig. No.	Height Above Bottom (m)	Average Velocity (cm/sec)	Max. Vel. (cm/sec) Up	Max. Vel. (cm/sec) Dwn	Average Velocity (cm/sec) Up	Average Velocity (cm/sec) Dwn	Average Velocity (cm/sec) Cross	Average Duration (hrs) Up	Average Duration (hrs) Dwn	Average Duration (hrs) Cross	Direction of Net Flow	T*	Avg. Cycle Length (hrs)	Tidal Range (ft)	Tidal Range (m)
	57-01-111B	1445	40, 119	30	6.7	22	23	8.8	8.9	2.3	3.2	3.4	2.0	none	98	11.6	6.5	2.0
Monterey	12-01-48	165	40, 41	3	10.3	26	21	11.7	10.8	8.5	0.8	0.7	1.5	up	72	3.9	5.5	1.7
	33-01-82A	155	40-43	3	7.8	31	29	9.2	10.3	4.0	1.2	1.7	1.0	down	88	8.8	4.7	1.5
	33-01-82B	155	40-43	30	6.3	20	21	8.5	6.7	3.7	2.3	1.5	0.8	up	88	8.8	4.7	1.5
	34-01-83	357	40-42, 44, 45	3	10.1	30	30	13.8	11.4	5.2	2.2	2.3	0.7	down	90	8.8	4.7	1.5
	35-01-84	384	40-42, 44, 45	3	10.3	22	27	12.1	13.1	5.6	2.7	2.1	0.8	up	88	8.0	4.7	1.5
	58-01-112A	1061	40-42, 44, 45	3	17.2	28	22	19.7	16.6	15.3	3.9	1.4	0.5	up	92	5.7	6.3	2.0
	58-01-112B	1061	40-42	30	18.7	30	26	20.3	26.0	9.8	4.7	4.3	0.4	up	92	6.5	6.3	2.0
	59-01-113A	1445	40-45	3	10.0	30	30	13.2	11.1	5.8	4.3	1.8	0.7	up	92	8.0	6.3	2.0
	59-01-113B	1445	40-43	30	9.3	31	28	13.6	10.0	4.2	4.8	2.9	0.8	up	92	8.7	6.3	2.0
San Clemente	122-01-184	1372	46-49	3	10.4	25	29	10.5	12.8	8.0	0.6	0.7	0.1	down	333	11.0	7.5	2.3
	123-01-185	1646	46, 47	3	11.1	26	32	16.0	11.8	5.4	0.9	0.9	0.1	down	80	12.2	6.4	2.0
	124-01-186	1756	46-49	3	6.5	24	21	6.6	6.5	6.3	0.5	0.3	0.4	up	333	19.0	7.5	2.3
Fraser	90-01-152B	85	50, 51, 53	30	7.0	22	35	6.6	10.0	4.5	4.6	4.4	3.1	down	576	10.0	13.5	4.2
	91-01-153A	220	50, 52, 53	3	10.8	17	30	8.5	15.0	9.0	3.3	5.1	3.2	down	141	12.5	10.1	3.1
	92-01-154	148	50	3	11.0	***	***	***	***	***	***	***	***	***	24	12.3	9.4	2.9
	125-01-187	60	50, 51	3	7.4	29	17	10.9	5.9	5.3	0.3	0.1	0.1	up	142	11.6	15.3	4.7
	126-01-188	120	50, 53	3	15.0	32	36	14.7	20.1	10.3	0.5	0.7	0.2	down	142	12.0	15.3	4.7
	127-01-189	195	50, 52, 53	3	17.3	32	35	16.9	21.8	13.3	0.6	1.1	0.3	down	148	12.7	15.3	4.7
Cabo San Lucas	13-01-50	216	4, 54, 56	3	2.3	7	7	2.2	2.2	2.6	0.4	0.7	0.5	down	113	2.0	4.4	1.4
	14-01-51	137	4, 54-56	3	1.9	8	18	2.6	2.8	0.4	0.3	0.6	0.3	down	76	1.3	4.4	1.4
	14-02-53	137	54	3	**	**	**	**	**	**	0.4	0.4	0.3	**	83	1.0	4.7	1.5
	14-03-55	137	54	3	2.6	3	7	2.5	3.0	2.2	0.4	0.6	0.8	down	83	1.3	6.0	1.9
	15-01-52	328	54	3	6.1	18	32	5.8	6.4	6.2	0.7	1.3	0.7	down	122	4.1	4.7	1.5
	15-02-54	328	4, 54-56	3	5.1	18	35	4.0	8.0	3.5	0.5	1.3	0.4	down	43	4.1	6.0	1.9
San Jose	66-01-121	1408	55	3	7.9	20	17	10.5	9.5	3.9	4.1	6.7	0.9	down	39	20.0	4.0	1.2
	67-01-122	1860	55	3	5.4	18	18	6.3	5.6	4.2	2.1	2.4	1.3	none	38	12.0	4.0	1.2
Rio Balsas	68-01-123	1290	57, 58, 60	3	5.8	20	14	7.5	6.2	3.7	3.1	3.9	1.2	none	63	10.2	2.1	0.7

	69-01-124	1904	57, 58, 60	3	7.1	21	21	7.4	8.5	5.4	3.0	6.5	2.6	down	162	12.7	2.3	0.7
	70-01-125	285	57, 61, 122	3	9.7	31	34	12.6	13.9	2.6	1.7	2.0	0.6	down	112	5.7	2.3	0.7
	70-01-126	384	57, 58	3	11.0	30	32	12.2	16.3	4.6	1.5	3.1	1.1	down	136	6.0	2.5	0.8
	72-01-127A	658	57-59	3	6.7	23	23	8.1	8.1	3.8	2.3	2.9	0.9	down	135	8.5	2.5	0.8
	72-01-127B	658	57, 59	30	5.1	20	22	6.8	7.0	1.6	1.8	2.9	1.1	down	135	8.0	2.5	0.8
Petacalco	73-01-128A	110	57-59	3	8.0	35	31	12.1	7.9	4.0	1.1	1.3	0.5	up	129	3.6	2.5	0.8
	73-01-128B	110	57, 59	30	6.3	20	34	7.8	7.7	3.4	1.7	1.2	0.9	up	129	4.0	2.5	0.8
Hydro-grapher	45-01-99	713	63-65, 68	3	15.5	30	33	16.2	23.6	6.9	5.0	5.2	0.7	down	304	13.0	5.0	1.6
	46-01-100A	512	63-66, 68	3	14.0	44	33	14.8	21.0	6.3	3.5	5.2	0.7	up	296	11.5	5.0	1.6
	46-01-100B	512	63	30	7.9	39	26	8.8	10.3	4.5	6.1	4.0	0.6	up	296	10.3	5.0	1.6
	47-01-101A	348	63, 64-69	3	16.6	39	52	18.3	24.2	7.4	3.1	4.6	0.7	down	288	11.5	5.0	1.6
	47-01-101B	348	63, 65, 67	30	14.5	37	37	19.9	15.6	8.0	3.8	4.2	1.1	none	288	12.0	5.0	1.6
Wilming-ton	48-01-102	914	74-76	3	7.2	20	21	7.9	9.1	4.6	5.0	4.4	1.7	up	469	11.8	5.8	1.8
Hudson	49-01-103	1222	70, 72, 73	3	6.4	17	23	7.9	7.1	4.3	3.5	5.2	1.2	down	483	13.1	5.6	1.7
	51-01-105A	1765	70, 72, 73	3	7.1	18	12	7.9	7.9	5.5	6.4	3.7	1.0	up	312	12.4	5.6	1.7
	51-01-105B	1765	70	30	5.6	16	17	5.0	5.5	6.4	2.2	2.1	1.7	none	312	12.4	5.6	1.7
Christian-sted	96-01-158A	49	84, 91	3	3.9	12	10	3.9	4.6	3.2	0.4	0.4	1.0	down	115	1.9	0.9	0.3
	96-01-158B	49	84, 91	11	2.9	10	7	3.1	3.0	2.5	0.7	0.6	0.8	up	52	1.9	0.9	0.3
	95-01-157	56	84	3	4.2	11	20	4.1	5.3	3.3	0.3	0.3	0.4	down	115	1.3	0.9	0.3
	118-01-180A	1050	90, 93, 94	3	3.2	12	15	3.2	4.9	1.6	0.5	0.5	0.2	down	288	3.2	0.8	0.2
	118-01-180B	1050	90	30	2.8	10	10	3.3	3.3	1.7	0.9	0.8	0.3	up	289	3.2	0.8	0.2
	119-01-181A	1765	90, 94-96	3	4.4	14	12	5.8	3.9	3.6	0.6	0.5	0.3	up	288	6.5	0.8	0.2
	120-01-182A	2525	90, 94-96	3	2.6	7	8	3.6	2.3	1.9	0.5	1.8	1.0	down	132	13.3	0.8	0.2
	120-01-182B	2525	90	30	2.6	7	8	3.2	3.1	1.5	3.5	5.1	1.1	down	140	10.1	0.8	0.2
	121-01-183	410	90, 92	3	2.1	7	4	1.9	2.5	2.0	0.7	0.5	0.3	none	234	4.8	0.6	0.2
Salt River	98-01-160	48	84-86	3	5.7	18	26	4.4	9.0	3.7	0.2	0.1	0.7	down	115	1.0	1.3	0.4
	99-01-161B	164	84-86	30	3.3	16	12	3.6	3.6	2.7	1.2	1.3	0.3	down	115	1.0	1.3	0.4
	115-01-177A	49	84, 87, 88	3	6.1	7	26	6.0	7.4	5.0	0.8	6.4	0.6	down	238	1.9	0.9	0.3
	115-01-177B	49	84, 87, 88	30	8.7	17	14	9.1	8.4	8.5	4.1	1.5	1.0	up	246	1.9	0.9	0.3
	116-01-178A	238	84, 89	3	2.0	7	8	2.2	1.9		0.2	0.2	0.1	up	211	1.1	0.9	0.3

*Total duration in hours. **Direction data only; no velocity data. ***Velocity data only; no direction data.

Canyon	st-lo-ser	Depth (m)	Fig. No.	Height Above Bottom (m)	Average Velocity (cm/sec)	Max. Vel. (cm/sec) Up	Max. Vel. (cm/sec) Dwn	Average Velocity (cm/sec) Up	Average Velocity (cm/sec) Dwn	Average Velocity (cm/sec) Cross	Average Duration (hrs) Up	Average Duration (hrs) Dwn	Average Duration (hrs) Cross	Direction of Net Flow	T*	Avg. Cycle Length (hrs)	Tidal Range (ft)	Tidal Range (m)
	116-01-178B	238	84	30	2.3	9	9	2.2	2.8	1.9	0.6	0.9	0.6	down	211	1.8	0.9	0.3
	117-01-179A	130	84, 89	3	1.7	12	13	1.5	2.5	1.0	0.3	0.2	0.1	down	121	2.5	0.9	0.3
	117-01-179B	130	84	30	4.6	12	15	4.4	5.4	4.0	1.0	1.4	0.8	down	121	3.4	0.9	0.3
Rio de la Plata	113-01-175	238	78-80, 124	3	11.9	48	68	8.6	20.0	7.2	0.8	0.9	0.2	down	238	2.8	1.7	0.5
	114-01-176	439	78, 80, 81, 83	3	6.7	28	32	8.5	7.4	4.3	1.0	1.4	0.5	down	392	3.5	1.7	0.5
Kauai	63-01-117A	290	97, 98	3	4.2	18	33	4.0	4.9	3.7	0.3	0.3	0.5	down	75	1.0	2.2	0.7
Nohili	53-01-107A	4206	97, 106	3	1.0	7	7	1.2	0.7	1.1	0.6	0.8	1.0	none	130	11.5	2.4	0.7
	53-01-107B	4206	97	30	0.9	4	4	0.8	0.6	1.3	0.4	0.7	0.9	none	130	11.5	2.4	0.7
	77-01-132	2213	97, 106	3	4.0	14	14	3.2	5.6	3.1	2.9	2.2	1.6	down	166	7.8	2.5	0.8
Kaula-kahi	54-01-108	3904	97, 103	3	3.0	7	9	3.4	2.7	2.9	6.7	2.4	2.4	up	138	15.0	2.4	0.7
	55-01-109A	3255	97	3	**	**	**	**	**	**	5.4	3.7	3.1	**	140	12.7	2.4	0.7
	55-01-109B	3255	97, 103	30	3.9	9	12	4.5	4.9	2.2	4.9	7.1	0.7	down	140	12.7	2.4	0.7
	75-01-130	1737	97, 101, 102, 104, 105	3	13.1	26	24	14.0	14.0	11.5	3.3	2.1	4.6	up	169	12.5	3.0	0.9
	76-01-131	1939	97, 101, 102, 104, 105	3	10.9	21	21	11.2	11.9	9.7	3.5	3.9	2.6	none	169	13.4	3.0	0.9
	62-01-116A	240	97-100	3	7.9	18	35	7.4	10.7	5.7	0.3	0.5	0.3	down	274	1.3	2.2	0.7
	65-01-119A	770	97-100	3	8.1	22	31	7.7	11.8	4.7	0.9	1.9	0.4	down	171	3.6	2.2	0.7
Abra	104-01-166A	622	107-109, 123	3	8.8	31	55	11.2	11.8	3.7	1.6	4.2	0.6	down	104	11.5	3.0	0.9
	104-01-166B	622	107-109, 123	30	7.4	31	54	10.5	9.6	2.2	1.7	2.4	0.7	down	104	11.5	3.0	0.9
Pinget	106-01-168B	1350	107, 110	30	8.1	37	24	10.6	8.4	5.3	4.0	2.1	1.3	up	88	10.0	5.6	1.7
Congo	24-01-69	400	111-113	3	5.7	22	13	6.6	5.9	4.5	3.3	3.1	0.8	up	89	12.0	4.1	1.3

*Total duration in hours. **Direction data only; no velocity data. ***Velocity data only; no direction data.

Table 2. Depths and Maximum Speeds of Up- and Downcanyon Currents at Each Station 3 m Above Bottom

Depth in Meters	Location	Sta. No.	Max. Speed (cm/sec)	
			Up	Down
40	La Jolla	100	*12	10
46	La Jolla	1	*32	16
49	Salt River	115	7	*26
49	Christiansted	96	*12	10
56	Christiansted	95	11	*20
60	Fraser	125	*29	17
95	Redondo	18	24	24
95	Redondo	18	*13	10
101	Newport	20	13	*18
107	La Jolla	2	*23	10
110	Rio Balsas	73	*35	31
120	Fraser	126	32	*36
130	Salt River	117	12	*13
137	San Lucas	14	8	*18
155	Monterey	33	*31	29
156	Carmel	11	15	*20
165	Monterey	12	*26	21
167	La Jolla	5	20	*21 avg
174	Hueneme	26	*26	23
190	Redondo	36	*33	10
195	Fraser	127	32	*35
201	Hueneme	84	*24	23
205	Carmel	31	28	*32
206	La Jolla	3	16	*23 avg
208	Santa Cruz	81	35	*38
216	San Lucas	13	7	7
220	Fraser	91	17	*30
238	Salt River	116	7	* 8
238	Rio de la Plata	113	48	*68
252	Newport	21	*13	10
256	Hueneme	85	24	*34
274	Redondo	19	*21	10
285	Rio Balsas	70	31	*34
290	Kauai	63	18	*33
328	San Lucas	15	18	*32
328	San Lucas	15	18	*35
348	Carmel	32	15	*36
348	Hydrographer	47	39	*52
357	Monterey	34	30	30
357	Santa Cruz	29	*36	24
362	Redondo	37	*30	27
362	Hueneme	78	30	*32
366	Newport	61	9	*11
366	Newport	61	*26	21
375	Hueneme	27	34	34
375	La Jolla	10	*19	18 avg
384	Monterey	35	22	*27
384	Rio Balsas	71	30	*32
406	Hueneme	86	31	*33
410	Christiansted	121	* 7	4
439	Rio de la Plata	114	28	*32
448	Hueneme	28	32	32
458	Santa Monica	23	27	*30
512	Hydrographer	46	*44	33
585	Santa Cruz	30	22	*29
622	Abra	104	31	*55
658	Rio Balsas	72	23	23
713	Hydrographer	45	30	*33
725	Wilmington	NOAA	*40	35
770	Hanakapiai	65	22	*31
860	Santa Cruz	83	23	*31
914	Wilmington	48	20	*21
1050	Christiansted	118	12	*15
1061	Monterey	58	*28	22
1070	Carmel	56	27	*31
1222	Hudson	49	17	*23
1290	Rio Balsas	68	*20	14
1372	San Clemente Rift	122	25	*29
1408	San Jose	66	*20	17
1445	Carmel	57	23	23
1445	Monterey	59	30	30
1646	San Clemente Rift	123	26	*32
1737	Kaulakai	75	*26	24
1756	San Clemente Rift	124	*24	21
1756	Christiansted	119	*14	12
1756	Hudson	51	*18	12
1860	San Jose	67	18	18
1904	Rio Balsas	69	21	21
1939	Kaulakai	76	21	21
2525	Christiansted	120	7	* 8
3255	Kaulakai	55	9	*12
3904	Kaulakai	54	7	* 9

* Indicates direction of maximum current velocity.

avg Indicates average of several records at same location taken at various times.

Part 3
Support Data

Additional Methods of Presenting Results

Methods of presenting results which are used only in Part III are considered here. The simplest of these methods is virtually to copy the curves shown in the current-meter tapes, which show the continuous variation of current directions (solid line Fig. 30) and superpose on this curve dashed lines which show the approximate upvalley and downvalley directions. Above these direction plots we superpose speed in cm/sec using time intervals from 5 to 15 minutes. The direction of the flows is shown according to the magnetic compass used in our current meters. In the example included here (Fig. 30), the directions are concentrated near those of the up- and downvalley axis directtons (here 250° and 70°).

Crosscanyon flows are indicated by plotting the vectors of speed multiplied by the cosine of the angle of divergence of the current direction, from a direction thought to be normal to the axial trend of the valley (Fig. 31). Ordinarily these plots show very low velocities, but there are a few notable exceptions where currents were found to be predominantly across the valley axes, and had considerable strength.

Another method is to plot each of the vectors in a series of concentric circles with dots showing direction, 0 to 360° true, and speed indicated by distance from a central point (Fig. 32). In all cases where two vectors meet the same point only one dot is shown, but the diagram gives a clear idea of the current direction in relation to the axis of the valley.

Another method shows the cumulative total movement of the current during the time of observation and is referred to as a *progressive vector diagram.* Each vector increment is added (incrementally) to show the relative movement of the water (Fig. 33).

Because it is well-known that internal waves move across the continental shelf producing a repetition of the wave pattern (LaFond, 1962), it may be expected that such a pattern will also advance along the axes of the valleys. To test this idea we made simultaneous time-velocity curves to adjacent stations along valley axes and superposed the resulting curves by moving one timewise in relation to its neighbor to look for a repetition of the pattern. With a good match, we can determine the possible direction in which these waves advance either up or down a valley (Fig. 34). Thus, in the illustration used here, waves show an approximate repetition 15 minutes later at the upcanyon station in La Jolla Canyon, suggesting that internal waves moved up the canyon during the time of the measurements. Sometimes the curves do not match at all so far as we can determine, and in other cases the match is of questionable value.

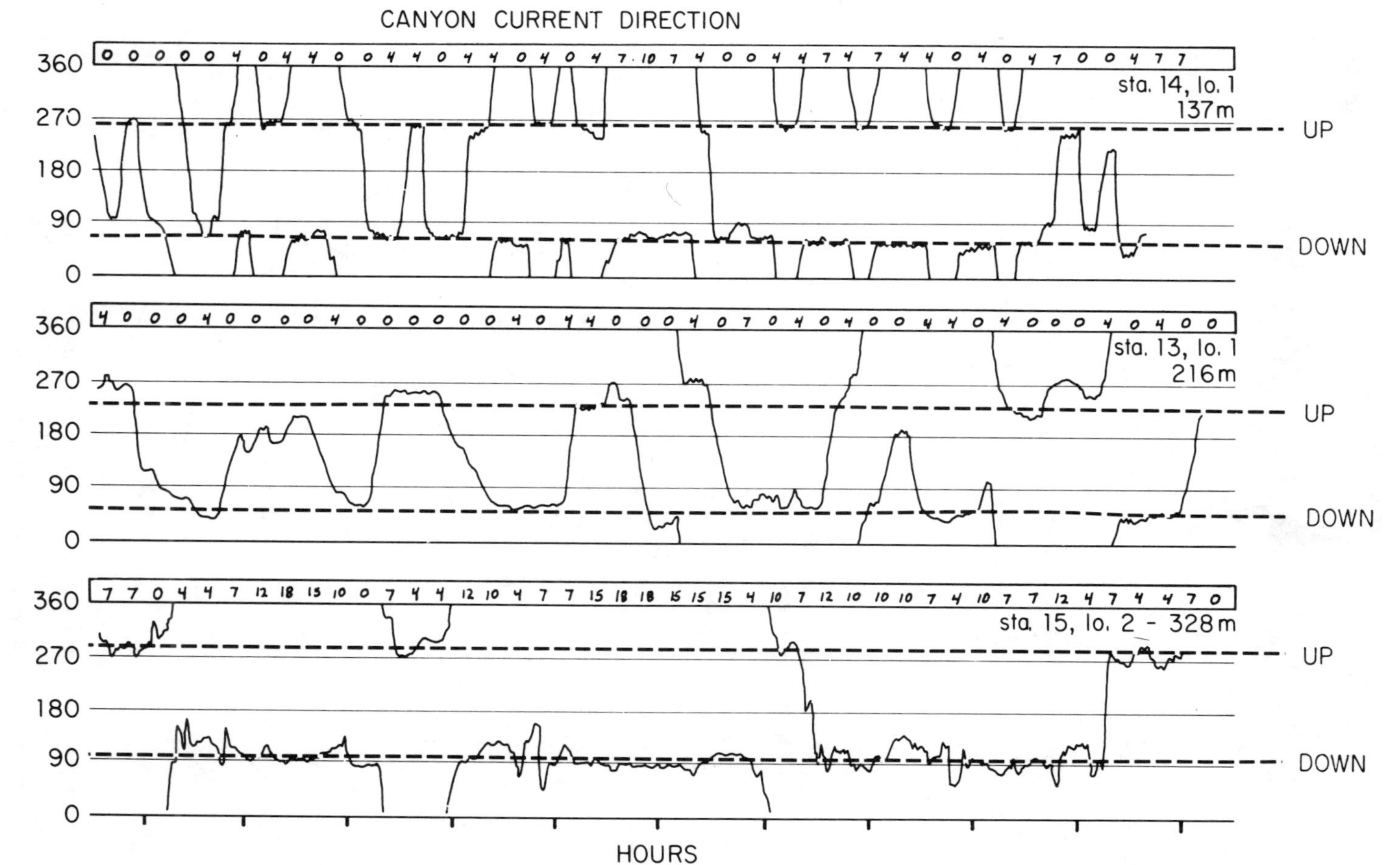

FIG. 30—Directions taken directly from current-meter record showing direction changes versus time and giving speeds (cm/sec) for each 15-minute interval (shorter units used in more recent records). The numbers represent average speed of the current for that period.

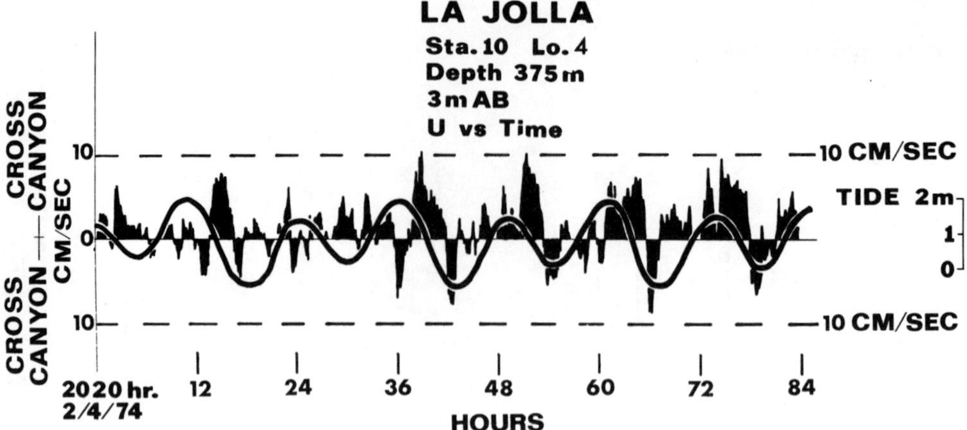

FIG. 31—Crosscanyon currents with speed corrected by the cosine of angle between line normal to axis and direction of current. Same record is used as in Figures 4C and 5 but note how much weaker currents were flowing across the canyon than along it and how much more frequent were the direction changes of crosscanyon currents.

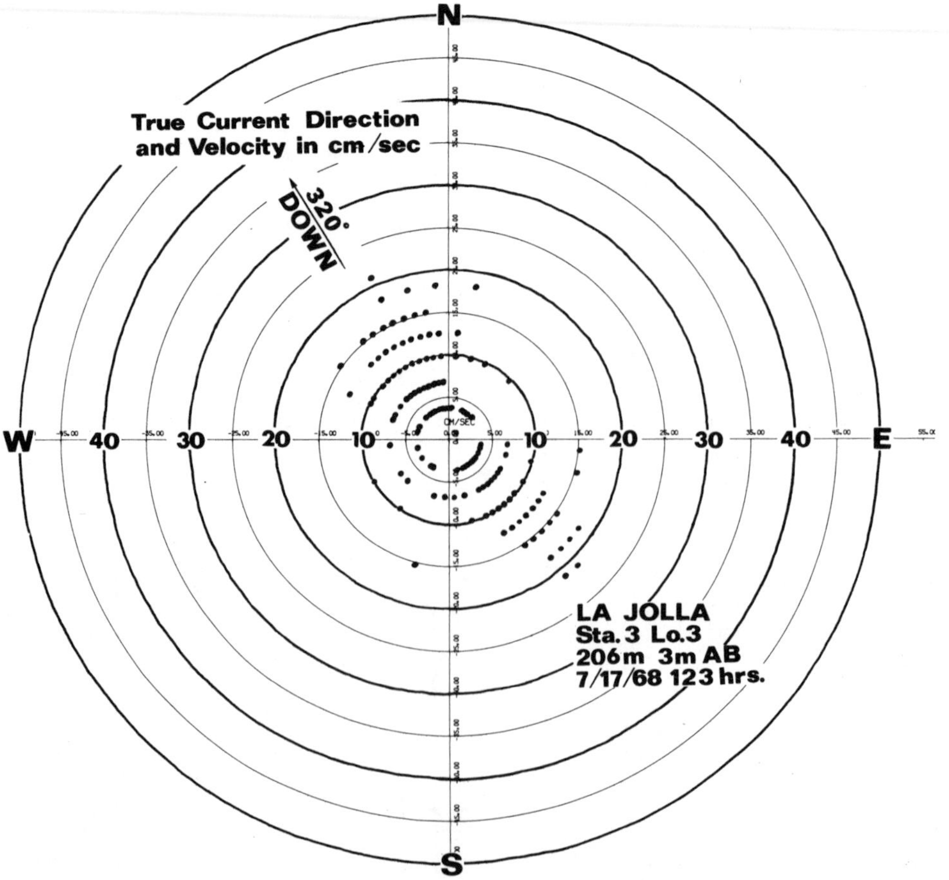

FIG. 32—Polar plot, dots represent current speeds and directions in this case to nearest 5° and nearest 2.5 cm/sec for each time division. The computer that develops these plots may hit the same spot more than once without indicating that this is the case. In La Jolla Canyon, currents were flowing approximately up and down the canyon axis which is northwest-southeast at this point.

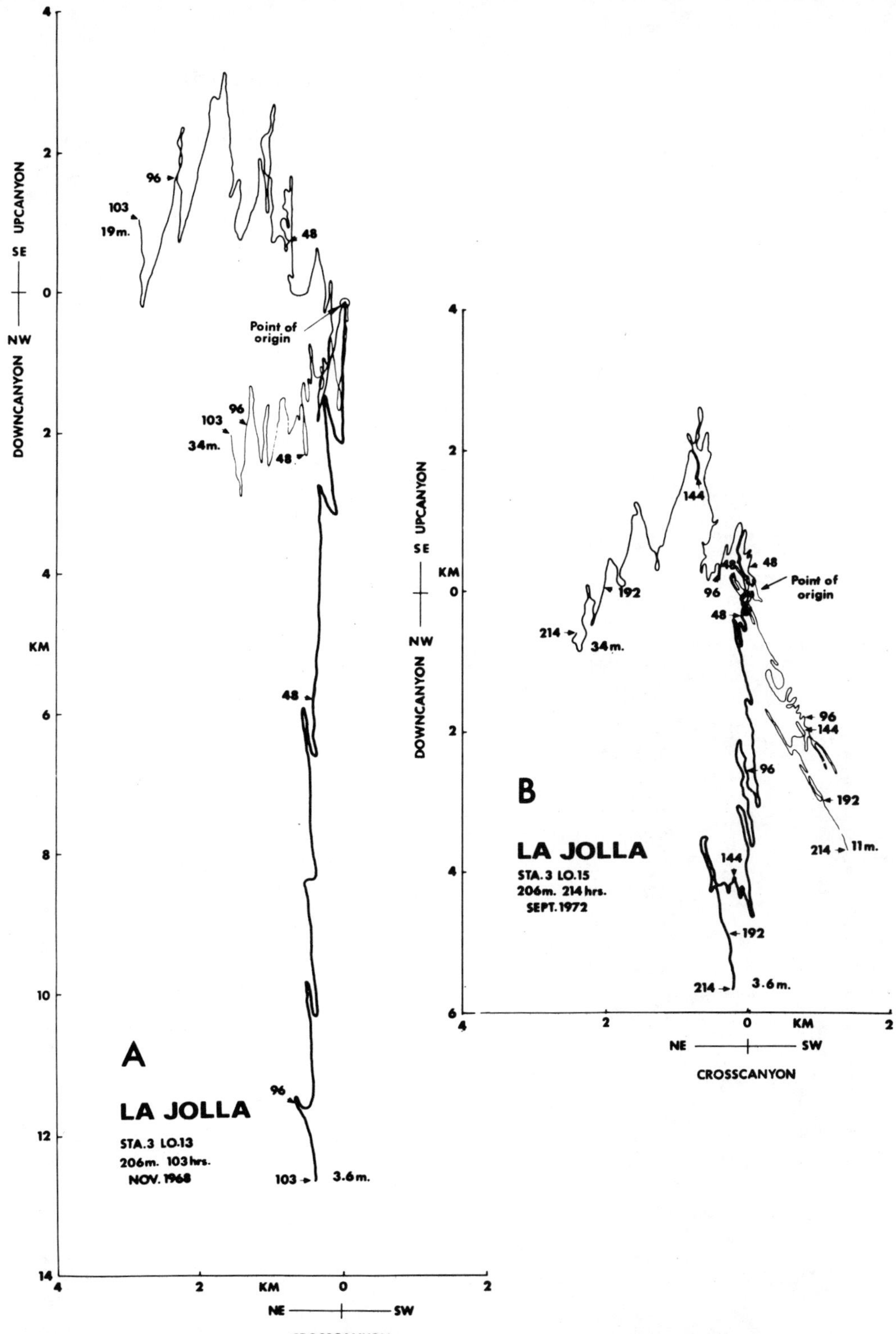

FIG. 33—Progressive-vector diagram shows cumulative advance of particles of water during the time of the record. Upcanyon is at the top of the page. Here where the current meter was 3 m above bottom the net flow was downcanyon interspersed with shorter upcanyon flows, whereas records from the meters at other heights above the bottom showed the flow as diagonally downcanyon or crosscanyon.

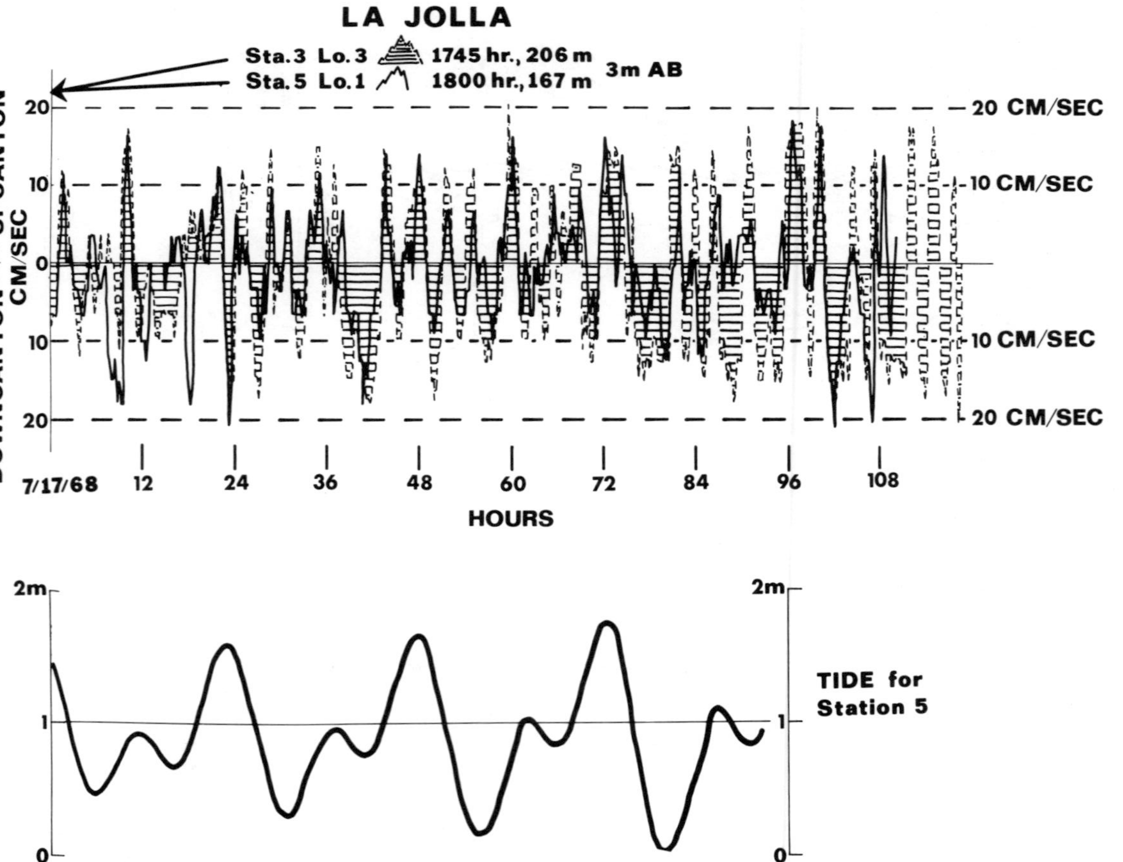

FIG. 34—Superposition of time-velocity curves from adjacent stations with time shift of one curve to show that best agreement between the two time-velocity curves. This method often indicates direction of advance of internal waves along the canyon axes. In this case the internal waves were moving up the axis. Lower curve shows tide for Station 5.

A method we used very little (but which may prove valuable to others) is to measure the duration of each upcanyon and each downcanyon flow at a series of stations where we take simultaneous recordings. Thus, in Figure 35 we measured the durations at stations in 78, 135, 205, and 375 m, plotting the depth as ordinate and the time as abscissa. This shows how the duration increases in length with depth of the axis.

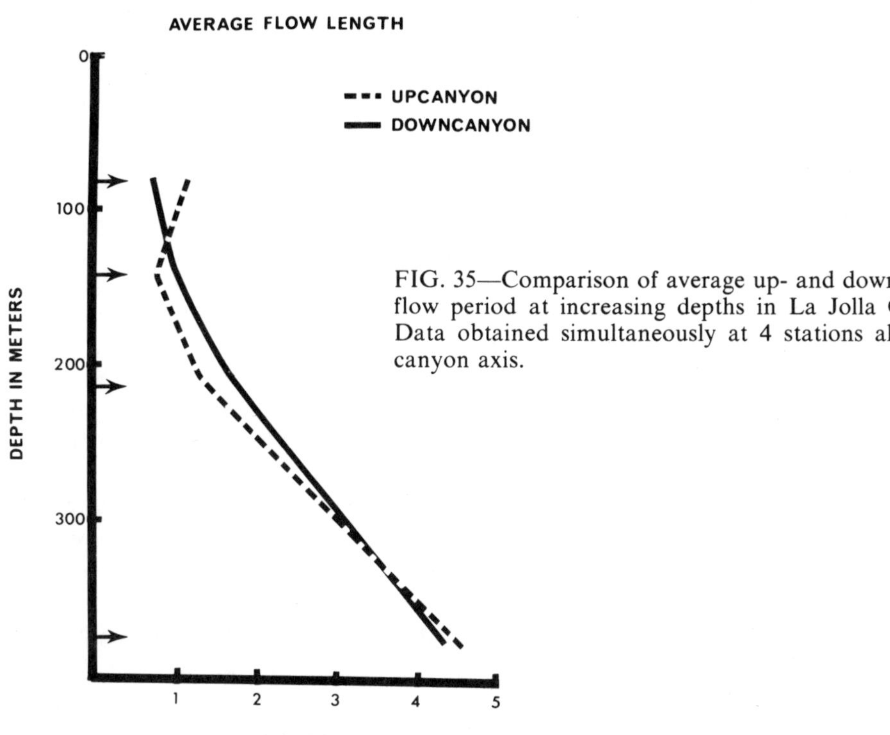

FIG. 35—Comparison of average up- and downcanyon flow period at increasing depths in La Jolla Canyon. Data obtained simultaneously at 4 stations along the canyon axis.

Because the time-velocity curves (series) can be mathematically broken down into individual sine waves, each of a specific period, amplitude, and phase which (when summed) approximate the originally measured series, we also can calculate a phase lead (or lag) of one time series with respect to another for adjacent stations. We can calculate the speed of advance of a wave and whether it was travelling up- or downcanyon, from the phase value of the period wave in question and the distance between stations. This method is not used extensively in this treatise, but will be given considerable attention by Linda Holmes in a forthcoming doctoral thesis.

Description of Currents

Using the described methods, currents from each of the areas are discussed in this section with illustrations so as to make available as much of our original data as possible. Because our records are, so far as we know, the most abundant obtained from the seafloor, we hope that other investigators will be able to use this information and make their own interpretations. We have no illusions that the interpretations we give here or in Part II represent all that can be drawn from these records and graphs. Also, more field work is certainly desirable to help future interpretations.

Currents in California Submarine Canyons

Most of the California coast is bordered by submarine canyons. These usually head close to the coast. Off southern California they extend seaward into the relatively shallow basins or troughs of the continental borderland, but off the central and northern part of the state they continue seaward well down the continental slopes to oceanic depths. Because of their convenient location near Scripps Institution of Oceanography, we conducted a large part of our investigations in these areas.

La Jolla and Scripps Canyons

Located off La Jolla near the Scripps Institution of Oceanography, La Jolla Canyon and neighboring Scripps Canyon (Fig. 36) were investigated more thoroughly than any other in the world. Scripps scientists and others have made extensive soundings and numerous dives into these canyons with scuba and deep-diving vehicles. Their general configuration is well known. Both of the canyons have heads close to the shore and extend seaward with relatively straight courses. They join almost at right angles at a depth of 300 m. Scripps Canyon is very narrow, and its gorge has extensive vertical or even overhanging walls and the floor is (in general) only a few meters across. La Jolla Canyon has a wider floor and less precipitous walls. Repetition of precise profiles has shown that the depths of these canyon heads are constantly changing (Shepard, 1951). They commonly start to fill rapidly with sediment but the process is interrupted by turbidity currents (Shepard and Marshall, 1973; Shepard et al, 1977), and (partly) by a slow creep (Dill, 1964). The character of the bottom is diametrically different from time to time, consisting alternately of sand, sandy mud, or kelp (Shepard et al, 1969). Rock floors apparently are exposed after turbidity currents.

Our current-meter studies were made mostly in La Jolla Canyon because Scripps Canyon proved to be a difficult place to acquire records, due to its narrowness and vertical walls. In many cases it appeared that our current meters were resting against the wall or some other obstruction that made the meter inaccurate or inoperable. Also, we lost three current meters during storms apparently due to turbidity currents and the great volume of kelp that moves down the canyon during these flows.

In the three shallow stations all four records had net upcanyon flow, but in the deeper stations, 21 records showed net downcanyon flow and only two had net upcanyon. Of the latter, one was the middle current meter in a three-instrument, vertical array, and in this case the bottom and top meters, respectively at 3 m and 34 m above bottom, had net downcanyon flow. Thus we only have one instance where net flows were upcanyon near the bottom in all the deeper La Jolla canyon stations.

The polar plots for La Jolla Canyon stations generally show good canyon direction control (Fig. 32). At the 206-m station many of the plots show the the fastest downcanyon currents were deflected somewhat to the north of axial direction (Fig. 32). This is probably related in some way to the contour of the canyon walls. The amount of crosscanyon flow is usually very small in La Jolla Canyon.

The frequency of up- and downcanyon alternations shows a relation to depth, but with some exceptions. Thus two lowerings at 46 m average 2.7 hours; at 78 m the

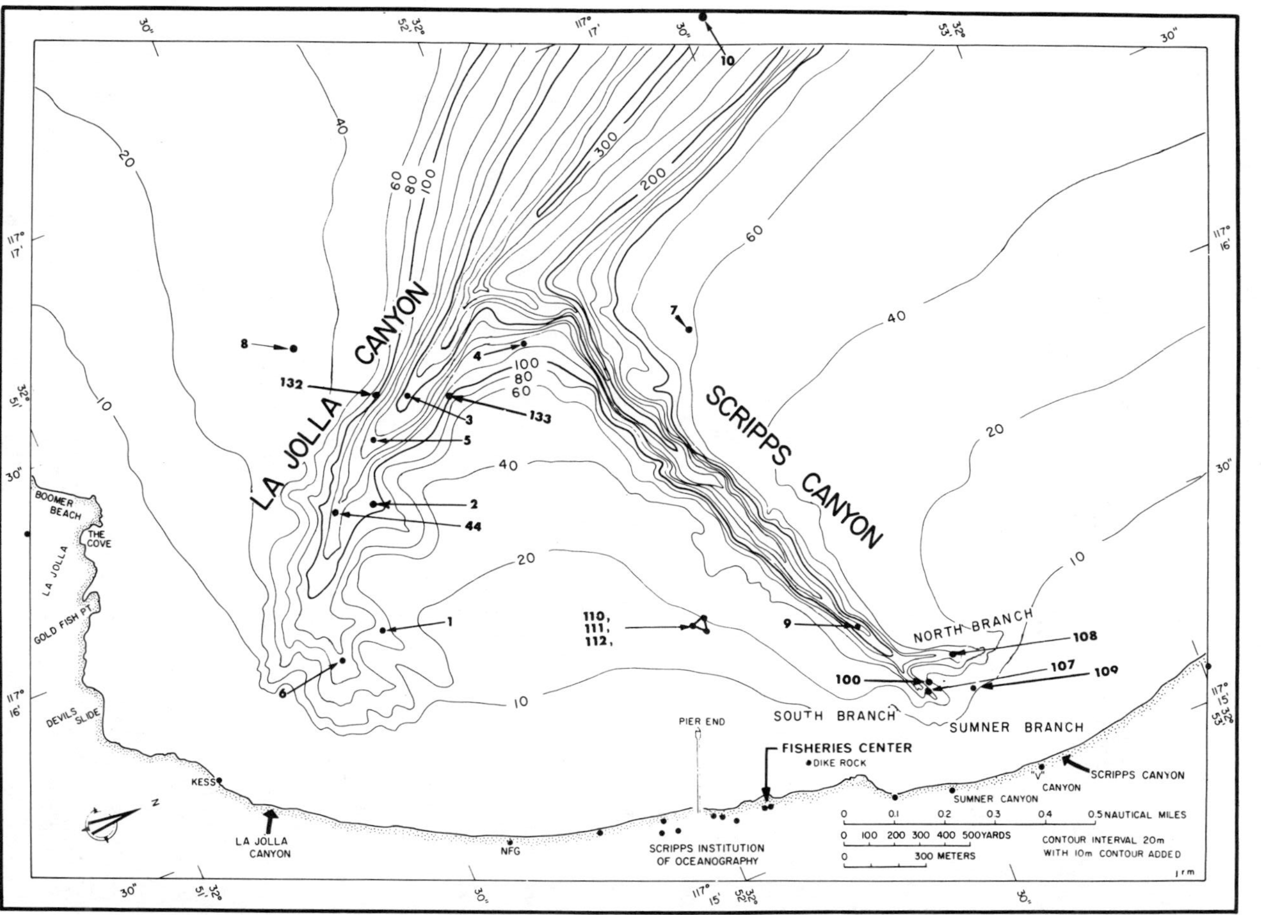

FIG. 36—Figure shows topography of La Jolla and Scripps Canyons. Locations of current-meter stations are indicated. Walls of Scripps Canyon are precipitous or even overhanging, whereas those of La Jolla Canyon are less steep with only small vertical portions. Topography based on surveys by Shepard.

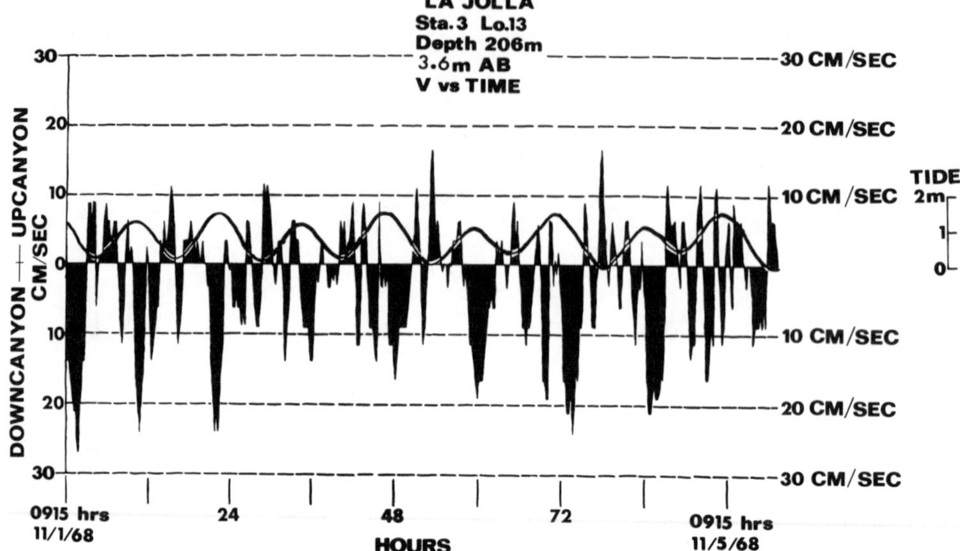

FIG. 37—Figure shows how a time-velocity curve may indicate a considerable tidal relation although the average length of up- and downcanyon oscillations is far more frequent than the 12.4-hour tidal cycle. Many of the peak flows are separated by approximately a 12.4-hour tidal interval. Such a relation is common at axial depths somewhat shoaler than where complete tidal relations are found.

two lowerings average 3.3 hours; at 107 m the one lowering averages 3.7 hours; at 135 m one lowering averages 2.7 hours; at 167 m the average of eight lowerings is 3.6 hours; at 207 m the average of 18 lowerings is 3.0 hours; at 375 m the average of four lowerings is 11.5 hours; and at 585 m the average of two lowerings is 12 hours. However, the 207-m station shows a considerable relationship to the tide (Fig. 37), and the 375-m station is almost perfectly tidal with a few added small-scale alternations (Fig. 4C). Thus the cycles become virtually tidal at axial depths of about 300 m. Comparison of the duration of upcanyon and downcanyon currents (Fig. 35) shows similar results.

Because Stations 3 and 5 are spaced only 600 m apart, it was possible to watch the progress of internal wave trains that move up along the canyon floor (Fig. 34). We obtained comparisons of the time-velocity curves at these two stations on six occasions, and four of them match well. These all show a later arrival of the waves at the shoaler station with an average difference of about 30 minutes. The other two have only weak correlation but indicate progress upcanyon. Comparing Stations 3 and 10, we also found arrivals later at the upcanyon station, in this case 3 hours and 40 minutes later at Station 3. They are separated by 1,800 m which makes the rate of advance upcanyon somewhat slower than that between Stations 3 and 5. Two other comparisons with poor fits also suggest upcanyon advance of the waves between Stations 3 and 6 and between 3 and 4. Thus there is little doubt that the train of internal waves moves up La Jolla Canyon.

On one occasion when we had current meters at 2 and 4 m above the bottom the records were very similar. Much less similarity developed when we had current meters at 3.6, 11, and 34 m, and on another occasion at 3.6, 19, and 34 m. The general character of the time-velocity curves shows a resemblance, especially in the 3.6- and 11-m comparison (Fig. 38), but there is only a general resemblance between the 3.6- and 34-m curves although most of the faster currents occur at approximately the same time in these comparisons. At the same station there is only a rough correspondance between 3 and 34 m (Fig. 6). La Jolla Canyon is somewhat atypical in this respect. Usually we found better agreement at the various levels above the bottom.

Scripps Canyon—On several occasions we obtained a record from Scripps Canyon which had such slow currents that we thought possibly the current meter was resting against the side wall or impeded in some manner from free motion. One record at an axial depth of 125 m shows currents up to 26 cm/sec with an alternation cycle of 2.4 hours. Another record at 128 m had currents up to only 12.5 cm/sec. Both of these records have net downcanyon flows that (in direction) closely correspond to the canyon axis. A current meter emplaced by scuba divers in Summer Branch of Scripps Canyon at a depth of 61 m recorded weak, irregular alternating currents not exceeding 11 cm/sec with reversal cycles averaging about two hours. Here the net upcanyon flows agree with those at the shallow stations in La Jolla Canyon.

Turbidity Currents—In Scripps and La Jolla Canyons the occurrence of turbidity currents was long suspected. Repeated surveys in both Scripps and La Jolla Canyons showed that the filling of the canyons is interrupted from time to time by episodes during which the water deepens appreciably (Shepard, 1951). Observers located in the National Marine Fisheries Services building on the cliffs above Scripps Canyon (see Fig. 36) have reported seeing boiling waves of mud rising to the surface from the canyon. Current meters placed in the head of Scripps Canyon have been carried away on several occasions. This was first observed by Inman (1970). They had a wire from their current meter conveying a record to the Scripps pier and recorded a current with a speed of 190 cm/sec preceding the loss of their current meter in the canyon head. In 1973 we placed current meters at 46 and 180 m in Scripps Canyon a few days before a storm and lost both of them. Previously in December of 1972, we placed two current meters in La Jolla Canyon at 2 and 4 m above the bottom at an axial depth of 206 m. A storm developed, and they did not surface at the designated release time. We were able to recover them later during a dive in the submersible *Nekton Gamma* (Shepard and Marshall, 1973); they were found in a mass of kelp at a locality 0.5 km downcanyon from where they were set. The records showed a rather fast downcanyon flow preceding a period of no current. Soon afterward, it stopped completely (Fig. 20). Although the velocity never exceeded 50 cm/sec at 2 m above the bottom and 30 cm/sec at 4 m, we suspect that this may have been the initial small pulse of a turbidity current, as it resembles examples observed later at other locations. The relation of this rapid downvalley pulse to the wind velocities is shown in Figure 20.

Newport Canyon

Newport Canyon heads close to Newport pier near where the sand spit protecting the harbor changes its trend (Fig. 39). The canyon extends in a general southerly direction diagonal to the coast. It has a depth only slightly below the surroundings and dies out in the fan to the south (Emery and Hülsemann, 1963; Felix and Gorsline, 1971). The canyon floor and walls are mostly mud covered, suggesting that weak currents predominate.

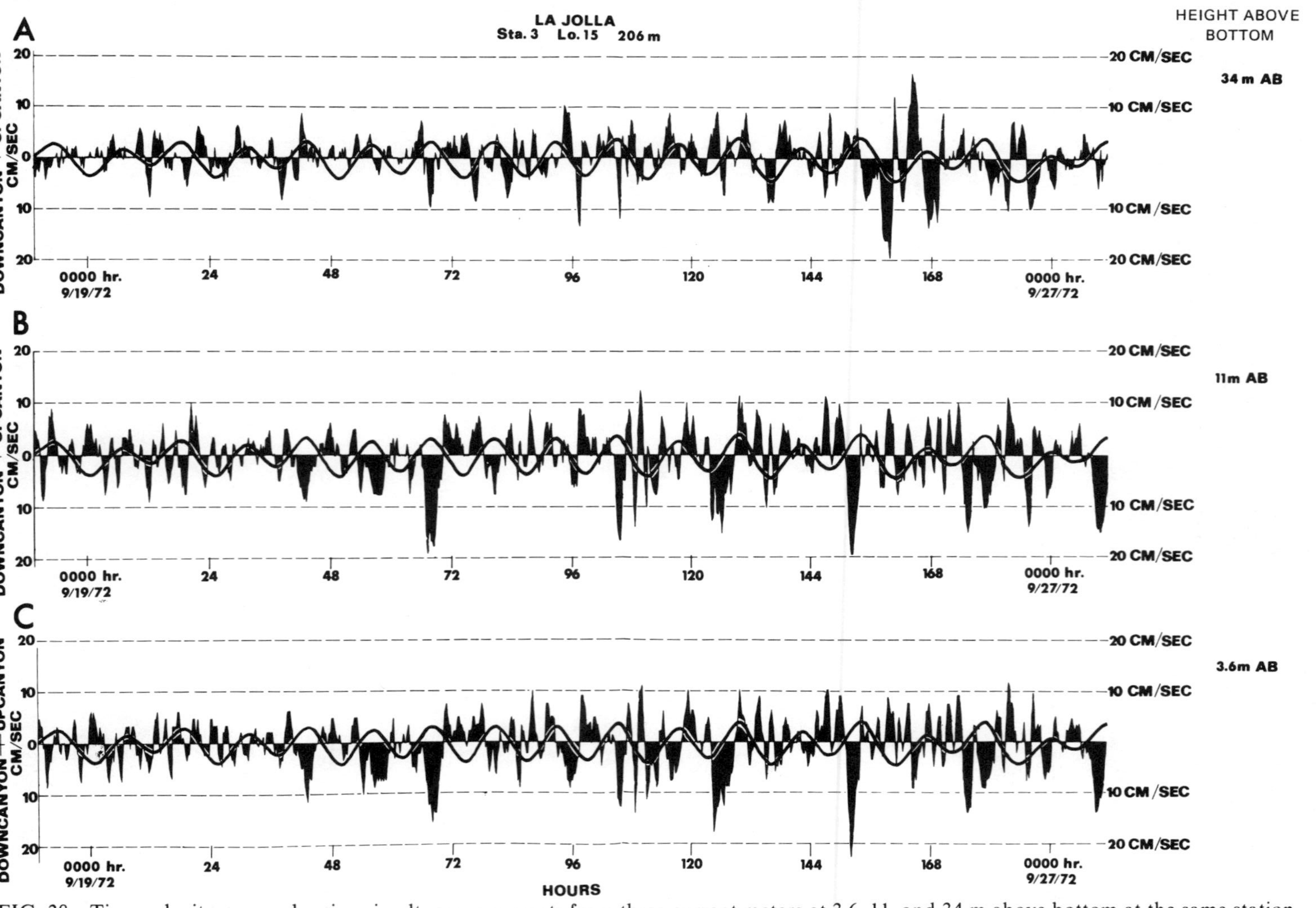

FIG. 38—Time-velocity curves showing simultaneous currents from three current meters at 3.6, 11, and 34 m above bottom at the same station in La Jolla Canyon at axial depth of 206 m. Resemblance applies only to some of the times of faster currents for the 11- and 34-m records but is much better for the 3.6- and 11-m records.

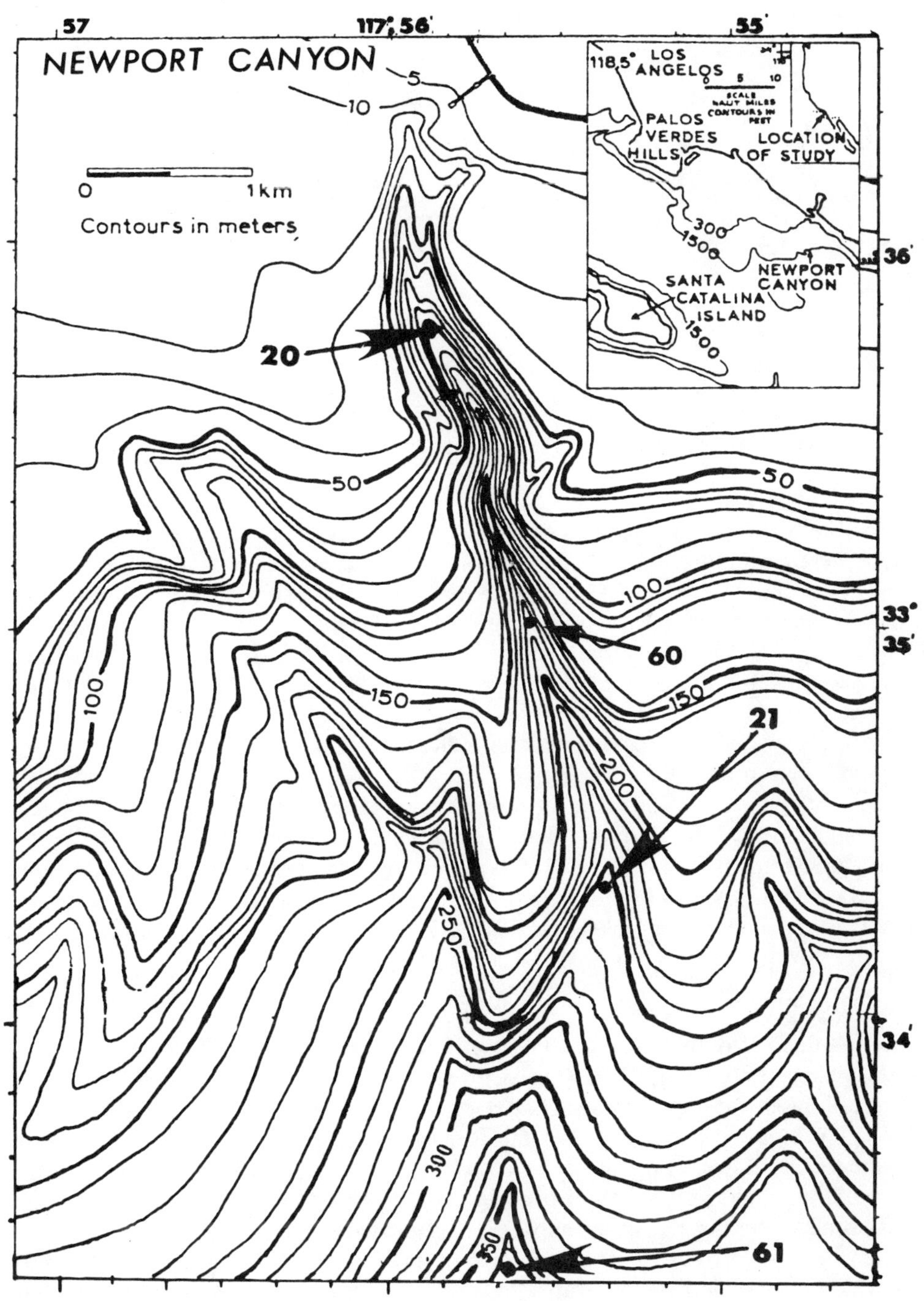

FIG. 39—Contour chart of Newport Canyon modified slightly from Felix and Gorsline (1971). Shows location of current-meter stations.

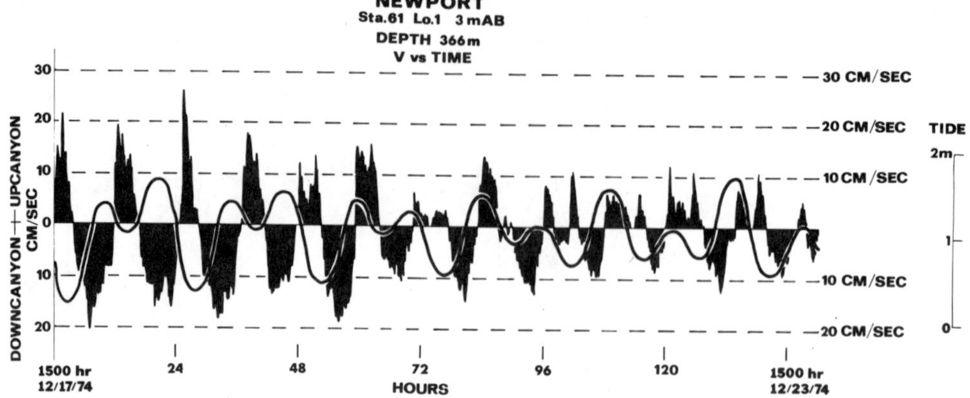

FIG. 40—Time-velocity plot of record from 366 m in Newport Canyon showing the only record from Newport Canyon with relatively fast currents. Notice close relation of time of current reversals to tides, but flow is in opposite direction from what tide would produce.

We obtained four current-meter records in the canyon; two at depths of 101 and 252 m were contemporaneous, and the others at 358 and 366 m were on different dates at Station 61. The most notable feature of the currents is that all but the one at 366 m had low speed in keeping with the mud bottom. The record obtained in December of 1974 had upcanyon velocities as much as 26 cm/sec, but the others are slower, mostly less than 10 cm/sec. The records at 101 and 366 m have net flow downcanyon, and those at 252 and 358 m have upcanyon net flow. The comparison between the two contemporaneous runs indicated that the pattern could be matched best, although poorly, by an upcanyon advance of internal waves.

The same depth relations to alternating cycles of up- and downcanyon flows was observed as in La Jolla Canyon. At 101 m the cycle averages 4.4 hours; at 252 m, 11 hours; at 358 m, 10 hours; and at 366 m, 11 hours. The deeper two are predominantly tidal with a few shorter periods that reduce the average (Fig. 40). At 358 m, it was suspected that the current meter landed on a terrace within the canyon. In general, currents flow alternating up or down the axis of Newport Canyon.

Redondo Canyon

Redondo Canyon is located at the southern end of Santa Monica Bay just north of the Palos Verdes Hills (Fig. 41). At present no river empties into the bay at this point although there is evidence that such a river existed in the past (Yerkes et al, 1967). The canyon lies directly off the Redondo pier and, according to U.S. Army Engineers, rapid deepening of the canyon head was a threat to the pier in 1920. The construction of the jetty north of the pier caused serious loss to shore-front property at Redondo because the usual supplies of sand from the north were blocked and because waves were converging just north of the canyon head (Shepard, 1977, p. 91). The canyon axis winds seaward in a westerly direction terminating in a fan in Santa Monica Basin at a depth of about 750 m. The transport of sediment along the canyon has been discussed by Beer and Gorsline (1971).

Six complete current-meter records have been obtained in this canyon, while two others recorded directions but no speeds. On one occasion we had three stations

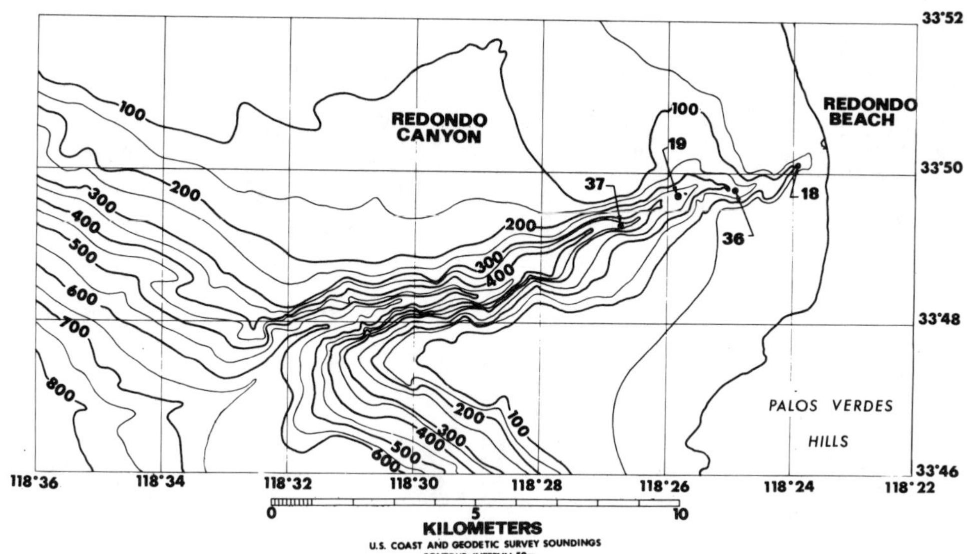

FIG. 41—Contour chart of Redondo Canyon and a portion of the fan and fan valley with its levees at the seaward end of the fan valley. Positions of current-meter stations are indicated (from U.S. Coast and Geodetic Survey soundings).

with five current meters, but at each station only one meter had a complete record. Depths ranged from 95 m to 365 m. The shallow station at 95 m has up- and downcanyon cycles averaging about 6 hours for this 21-day run (Fig. 42). Most of the faster upcanyon flows and some of the downcanyon flows agreed with the diurnal tide (Fig. 42). The current meter that operated at 280 m was 90 m above the bottom, but the current direction appears to correspond mostly to the canyon axis. At 190 m the record shows a fairly good correlation with the diurnal tides but has average cycles of 5.6 hour length (Fig. 43). At 362 m the up and down alternations are irregularly spaced averaging about 7.4 hours but have some relationship to diurnal tides. It is unusual for California canyons not to show good tidal relations at this depth. The very small neap tides occurring at this time may account for this discrepancy.

The current meters in Redondo Canyon had three records with net upcanyon and three with net downcanyon flows. The flow at 182-m depth is decidedly upcanyon (Fig. 44). Most of the currents are less than 20 cm/sec, and the highest reading is 33 cm/sec in an upcanyon direction.

Of the two cases where plots with contemporaneous records can be compared at adjacent stations, only one pair can be matched and that showed many discrepancies (Fig. 44). In this case the internal waves appear to be moving downcanyon, unlike most other comparisons from canyons that extend in close to the California coast.

Santa Monica Canyon

Santa Monica Canyon heads 9 km offshore in Santa Monica Bay thereby differing from most southern California canyons in not heading near the coast. We obtained two records but the one at 250 m did not record speed, so our only complete

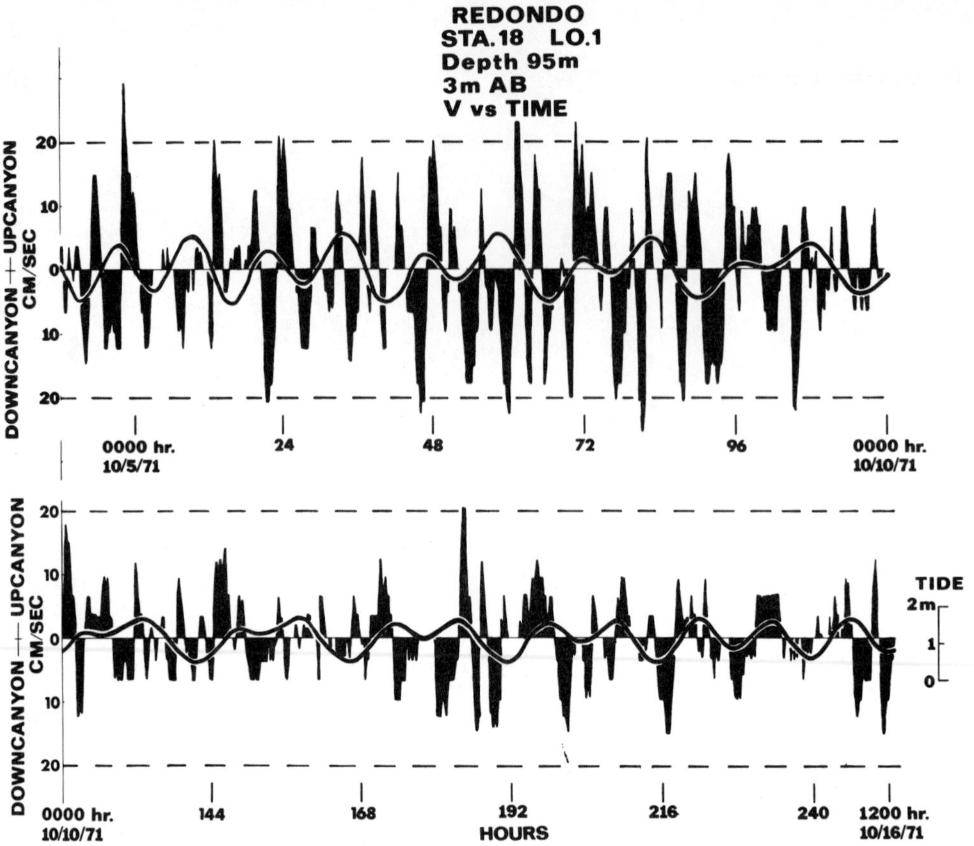

REDONDO
STA. 18 LO. 1
Depth 95m
3m AB
V vs TIME

DOWNCANYON — UPCANYON CM/SEC
20 10 0 10 20
0000 hr. 10/5/71 24 48 72 96 0000 hr. 10/10/71

DOWNCANYON — UPCANYON CM/SEC
20 10 0 10 20
TIDE 2m 1 0
0000 hr. 10/10/71 144 168 192 216 240 1200 hr. 10/16/71
HOURS

FIG. 42—Portion of the 21-day record at 95-m axial depth in Redondo Canyon. First six days show that the fastest upcanyon flows, and some of the fastest downcanyon flows are clearly separated by semidiurnal intervals and hence related to the tides despite the shallow depth. Little tidal relation is found in the subsequent 18 days of the record.

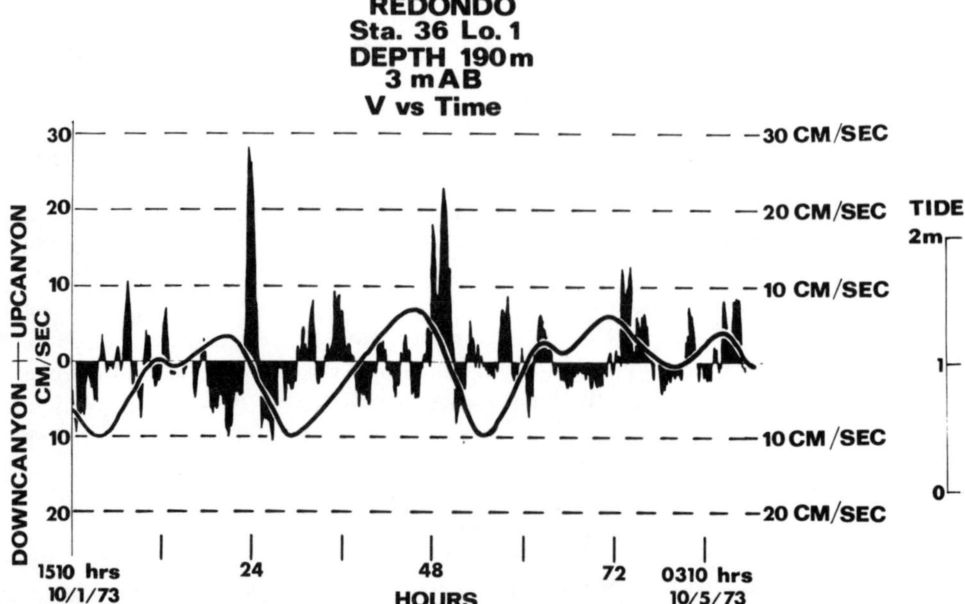

FIG. 43—Possible diurnal tidal relation at 190-m depth of two major upcanyon peak flows in Redondo Canyon during a neap tide period. Average alternation cycles are much shorter.

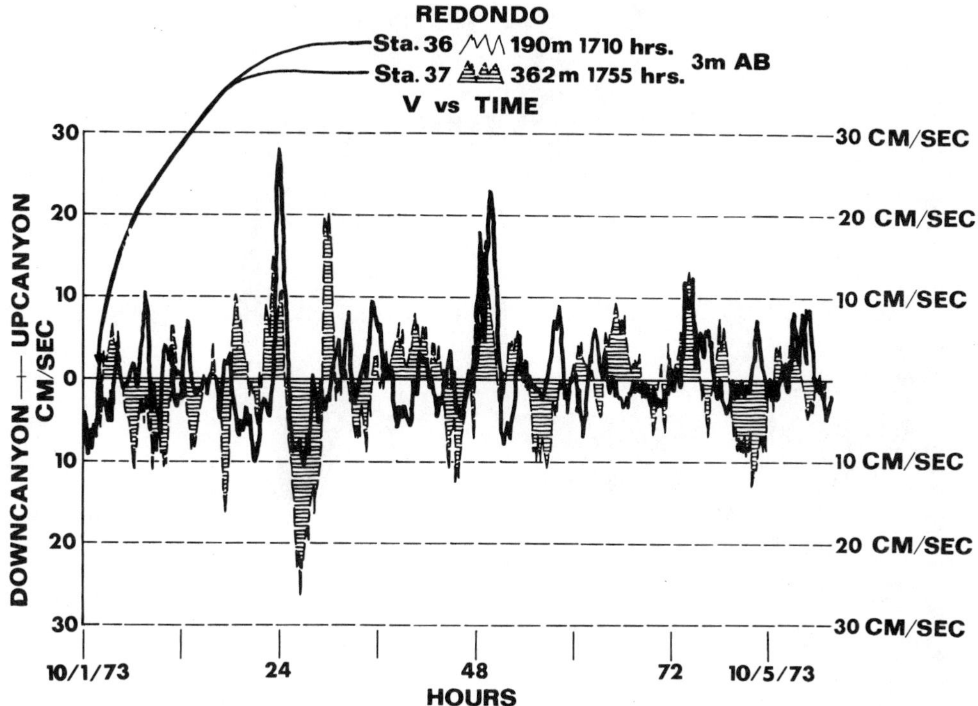

FIG. 44—Superposition of time-velocity curves from 190- and 367-m axial depths. The curve for the deeper station has been shifted so that its times are 45 minutes later than for the shallow station making the best of a poor fit. This may indicate a downcanyon advance of the internal waves.

record was at 458 m. The direction changes recorded at 250 m indicate that part of the time a semidiurnal tidal relationship existed while the rest of the time a higher reversal frequency occurred that was not related to the tides. This agrees well with what we have found elsewhere along the California coast at similar depths.

At 458 m the time-velocity curve (Fig. 45) shows good agreement with the tide, except that the faster currents apparently are related to the smaller tidal changes rather than to the large ones as is more generally true. The net flow is downcanyon and the fastest currents, about 30 cm/sec, are in that direction. The polar plot shows flow directions close to the canyon axis for the most part.

Although this one short record may not by typical, it does provide a suggestion that the effluent from the Los Angeles sewer outfall, which is near the head of Santa Monica Canyon, is transported down this canyon, but more information would be desirable. The shallow station with only directions showed longer periods of downcanyon flow than upcanyon, further supporting the idea that the effluents are carried mostly downcanyon.

Hueneme Canyon

Hueneme Canyon extends south from the apex of the Santa Clara Delta at Point Hueneme. It heads within 200 m of the coast (again, as at Newport) at a coastal bend (Fig. 46A). A harbor has been developed at this point by dredging inside the

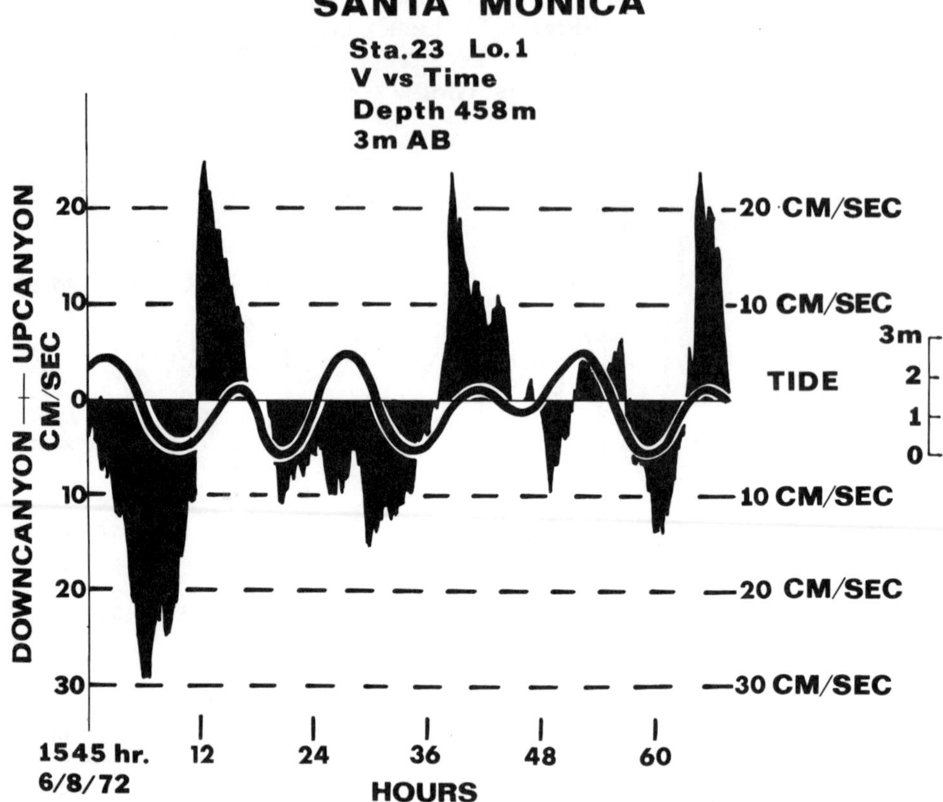

FIG. 45—Time-velocity curve for Santa Monica Canyon at a depth of 458 m at 3 m above bottom showing relations to both diurnal and semidiurnal tides.

delta and constructing jetties outside. The canyon winds seaward in a general southerly direction for about 21 km terminating in the Santa Monica Basin. Profiles normal to the canyon indicate that it is relatively flat-floored compared to other submarine canyons in the southern California area. At present there is no stream entering at Hueneme. So far as we can determine during historical times, the Santa Clara River always emptied about 11 km to the northwest of the canyon head. However, longshore currents are known to carry sediment from the Santa Clara River to the head of Hueneme Canyon (Drake and Gorsline, 1973).

We measured currents on four occasions in Huneme Canyon. Twice we had current meters at three locations along the canyon axis for four days and we have one single record of 22 days.

In February of 1973 we obtained four-day records at 174, 375, and 448 m. They were acquired during a period of strong westerly winds. In the mid-depth station the record was quite unique in that the fastest currents showed a wide divergence in direction from that of the canyon axis (Fig. 47). The polar plot shows that there were frequent currents between 20 and 30 cm/sec in directions varying from 280° to 110°, and the fastest currents, between 30 and 40 cm/sec, occur near the easterly end of this large arc with no currents of low speeds in this general direction. The strong easterly currents were flowing directly across the axis and they flowed in this direction for as much as 4 hours as shown by the crosscanyon flow plot (Fig. 48). These flows may have been related to the strong westerly or northwesterly winds that occurred during this period. It is perhaps significant that the times of strong crosscanyon flow were predominantly during periods of low tide and hence these periods are spaced at rather regular intervals. The stations at 174 and 488 m also show the effect of the crosscanyon flow in their progressive vector diagrams (Fig. 49). This suggests that a general outflow of water from Santa Barbara Basin may even penetrate to the floor of the canyon during some periods of strong westerly winds.

The relation of cycles of up- and downcanyon flow to tidal periods is similar to that of other stations in California canyons. At 174 m the average period is about 5 hours; at 201 m the average is 7.4 hours; at 256 m the average is 6 hours; at 375 m the average 11.7 hours; at 406 m it averages 11.5 hours; and at 448 m the average is 11.7 hours. Most of the cycles in the three deepest stations are clearly related to the tide (Fig. 50).

Comparing the time-velocity curves on the two occasions where we had three current meters at different depths in the canyon, we are able to match the curves but with some uncertainty (Fig. 51). These matches all indicate that the internal waves are advancing up the canyon. Thus the internal waves in Hueneme Canyon are comparable to those of most of the other California canyons.

The current speeds in Hueneme Canyon are similar to those found elsewhere in southern California; maxima are of the order of 30 to 40 cm/sec, flow being of a somewhat higher speed downcanyon in most but not all of the records. Also the new flow shown in progressive-vector diagrams is usually downcanyon. Thus it is only the surprisingly fast crosscanyon flows during February, 1973, operations that are atypical.

In Hueneme Canyon the only comparison between 3 and 30 m above the bottom was made at 256 m. The time-velocity plots are similar in this case (Fig. 52), and both had net downcanyon flow and speeds that were rather similar at the two heights.

The 22-day record at 362 m shows examples where the usual period was occasionally interrupted by a few more frequent cycles. Most (but not all) of these were during neap tides or periods when the heights of the two daily tides were quite different. The spread in direction of the downcanyon flows was noticeable as in several other Hueneme records.

Santa Cruz Canyon

Santa Cruz Canyon (Figs. 46B and 53) heads in the passageway between Santa Cruz and Santa Rosa Islands. Although the canyon trends generally to the southeast, there are two large bends along its course so that a part of the canyon trends northeast before resuming the southeasterly alignment. The surveys of this canyon indicate that the lower limit of the canyon and fan valley is well above the floor of the 2,000-m deep Santa Cruz Basin.

We investigated the currents in Santa Cruz Canyons on two occasions. The first was in February of 1973 at axial depths of 357 and 585 m. The second occasion was in September of 1975 with stations at axial depths of 208, 384, and 860 m although the midstation only recorded direction.

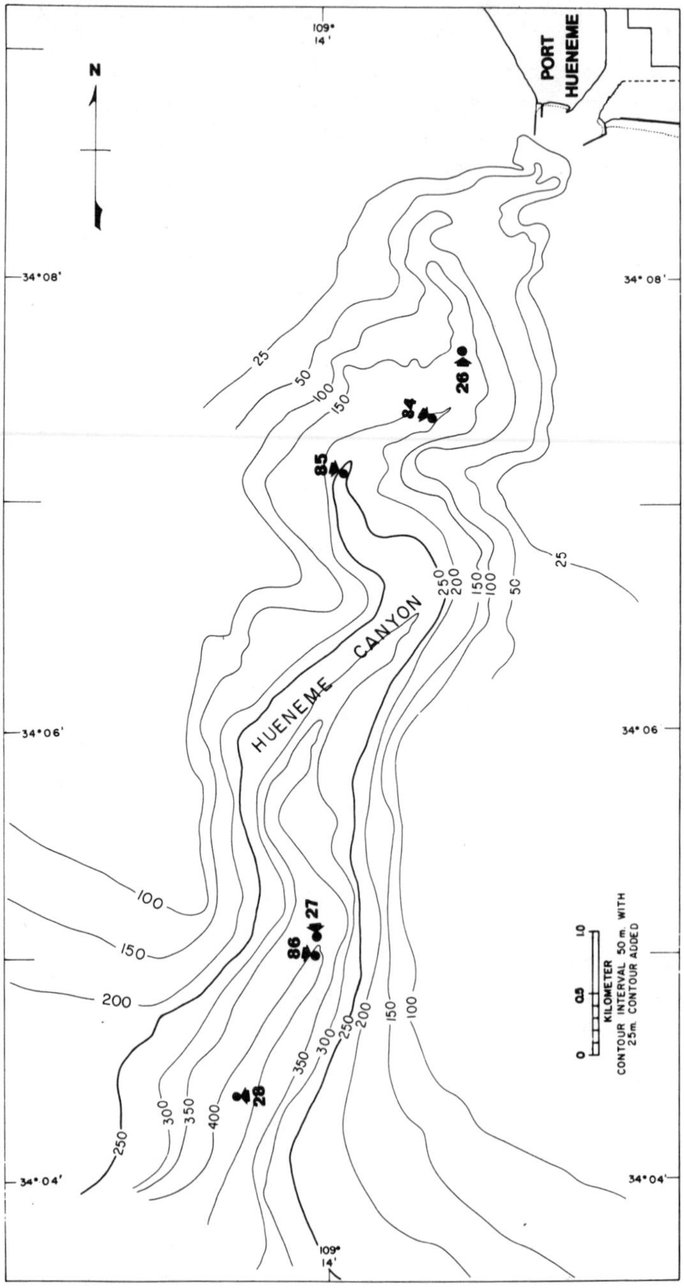

PORT HUENEME
N
109°
14'
34°08'
34°06'
34°04'
HUENEME CANYON
25
50
100
150
200
250
300
350
400
26
84
85
86
27
28
0 0.5 1.0
KILOMETER
CONTOUR INTERVAL 50 m. WITH
25m. CONTOUR ADDED

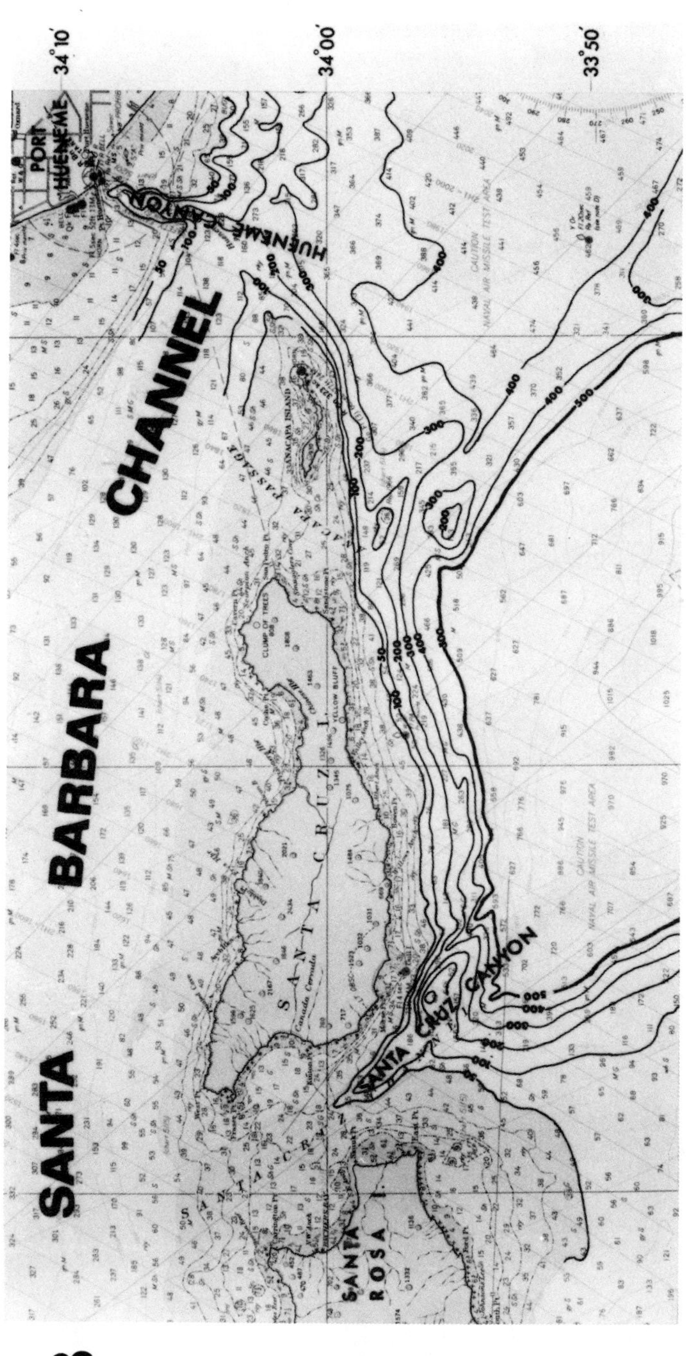

FIG. 46—**A**, the winding course of Hueneme Canyon with its relatively gentle walls. Located off the apex of the Santa Clara delta. Current-meter stations are indicated. **B**, showing relation of Santa Barbara Channel to Hueneme and Santa Cruz Canyons.

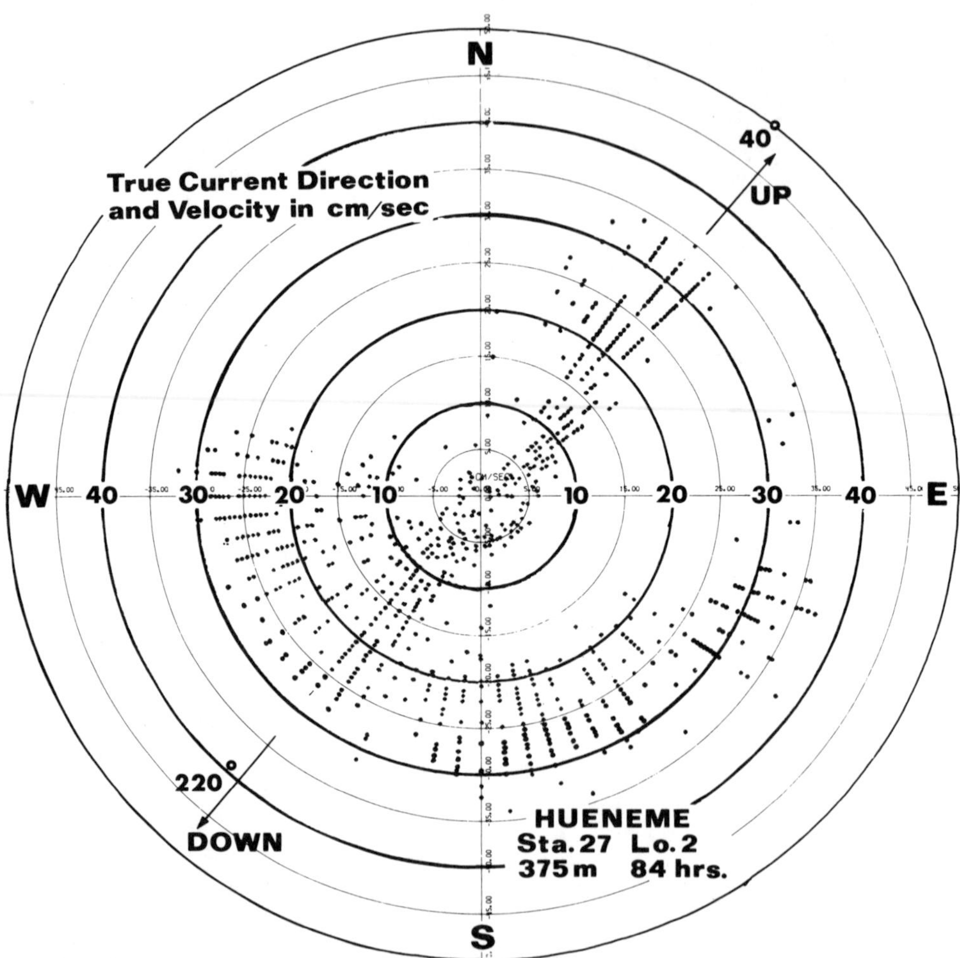

FIG. 47—Polar plot of currents in Hueneme Canyon at 375-m depth. A concentration of current in the upcanyon direction is shown, but with widespread currents on either side of the downcanyon axial trend. Fastest flows are in southeasterly crosscanyon direction. There are no slow currents associated with the southeasterly to easterly flowing fast currents. Strong winds from the northwest during the interval of the record may have contributed to these crosscanyon flows.

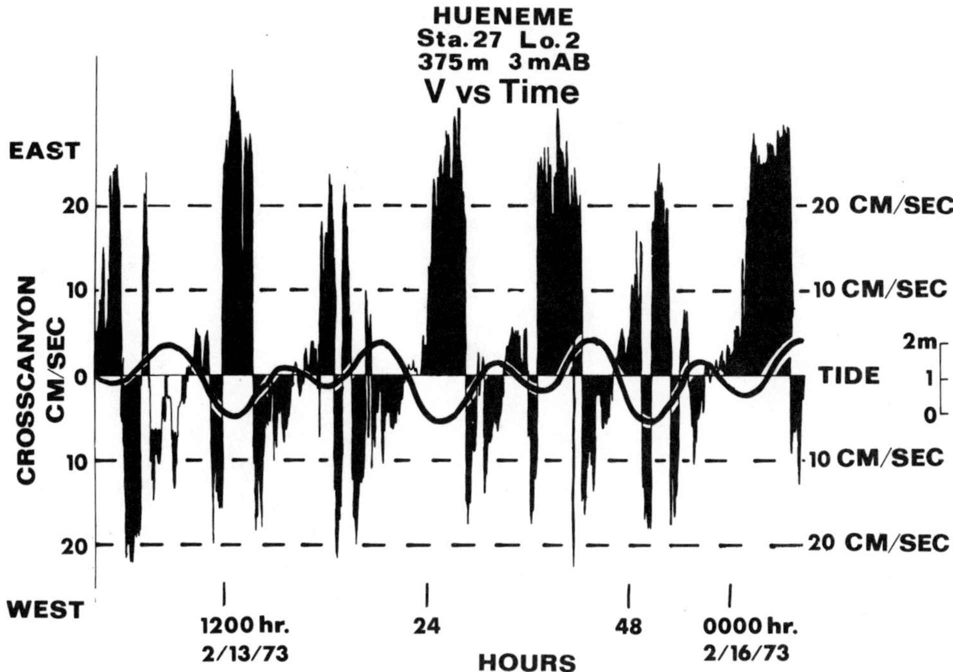

FIG. 48—Crosscanyon vectors of the 375-m record shown in Figure 47. Easterly high speed flows last for long periods compared to the westerly flows.

One of the peculiar features about the currents in Santa Cruz Canyon is that on both occasions the deepest station had approximately the same up- and down-canyon reversal cycle length as the shoalest. Thus at 208 m the average length is 5.8 hours; at 357 m it is 9.5 hours; at 585 m it is 10 hours; but at 860 m it is only 5.8 hours. It is notable that the periods were almost the same at the two different depths on both of our visits to the area. The station at 357 m showed a clear relationship to diurnal tides (Fig. 14), whereas at 585 m the cycles were both diurnal and semidiurnal (Fig. 54A).

In February, 1973, we obtained records during strong wind conditions: the first two days from the south and the last two from the more usual northwest. At the shallow station we had current meters at 3 and 29 m above the bottom, and at only 3 m above bottom for the deeper station. The faster flows occurred during the time of high and low tides. At 357 m the curves for 3 m and 29 m are very similar; both show rapid buildup of upcanyon velocity during the rising tide and slower fall off during high tide (Fig. 14). The progressive-vector diagram of the current shows net downcanyon flow, but there appears to be some relationship to the changing wind which was from the south at 14 to 22 knots during the first half of the record and from the northwest at 18 to 25 knots during the last half (Fig. 55). The south wind appears to have produced some upcanyon net flow in the early part of the record, and the subsequent northwest wind was accompanied by a decided downcanyon flow.

In the 1973 records it also is notable that the upcanyon flow patterns occurred later at the deeper station. A comparison of the two time-velocity curves shows that

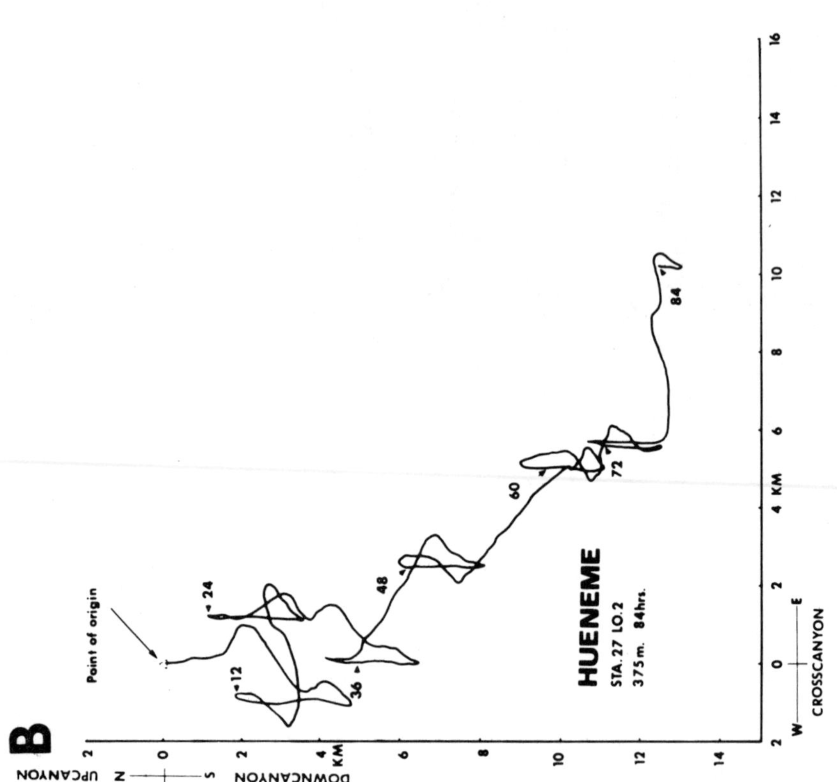
B
HUENEME
STA. 27 LO.2
375 m. 84hrs.
Point of origin
UPCANYON
DOWNCANYON
CROSSCANYON
KM

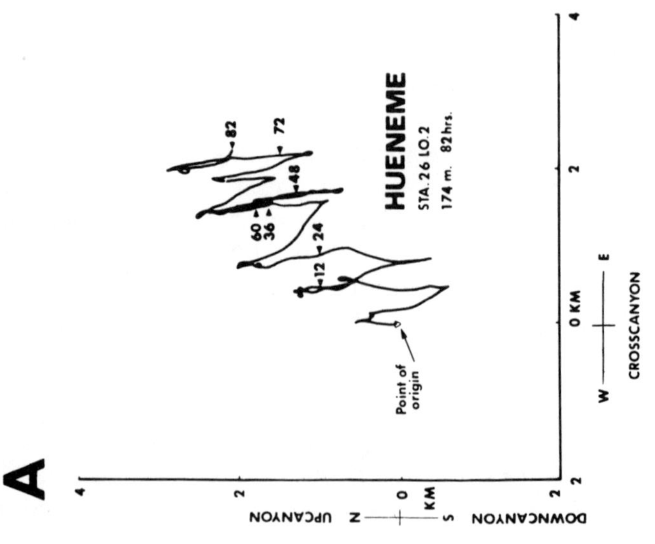
A
HUENEME
STA. 26 LO.2
174 m. 82 hrs.
Point of origin
UPCANYON
DOWNCANYON
CROSSCANYON
KM

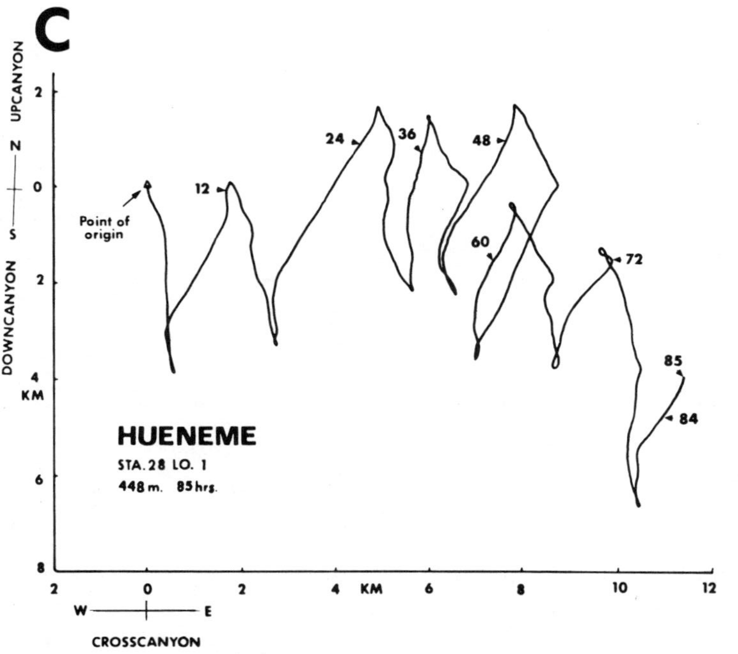

FIG. 49—Progressive-vector diagrams for the three stations occupied in Hueneme Canyon at a time of strong northwest winds, including the station shown in Figures 47 and 48. All of these show a net flow in an easterly direction but the shallow station shows net upcanyon flow whereas the deeper ones show net downcanyon flow. The effect of the northwest wind seems to be indicated at all depths.

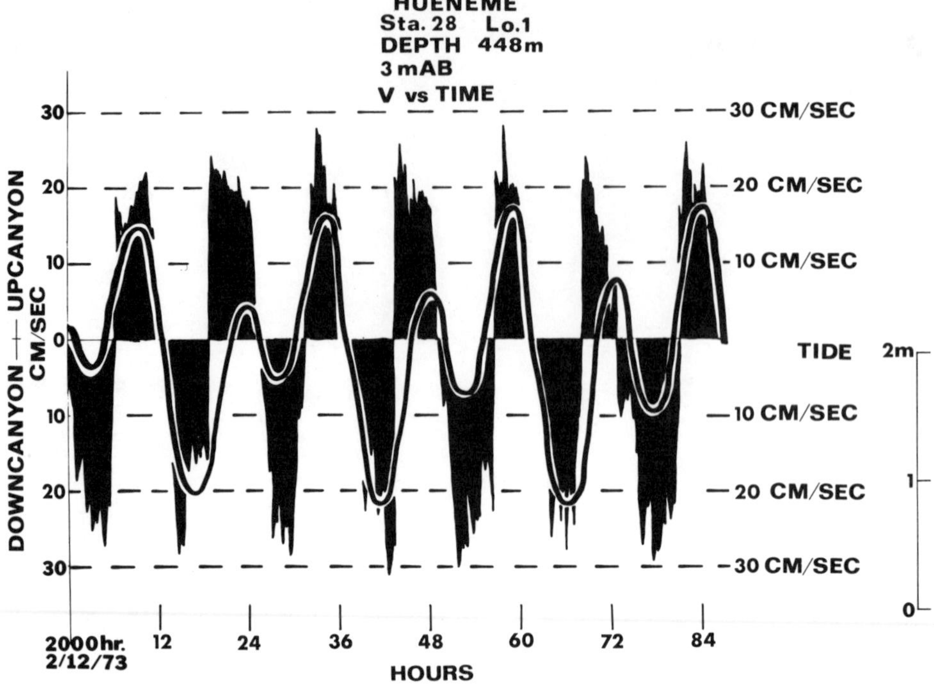

FIG. 50—Showing the clear relation of the up- and downcanyon current to the tide in Hueneme Canyon at 448 m.

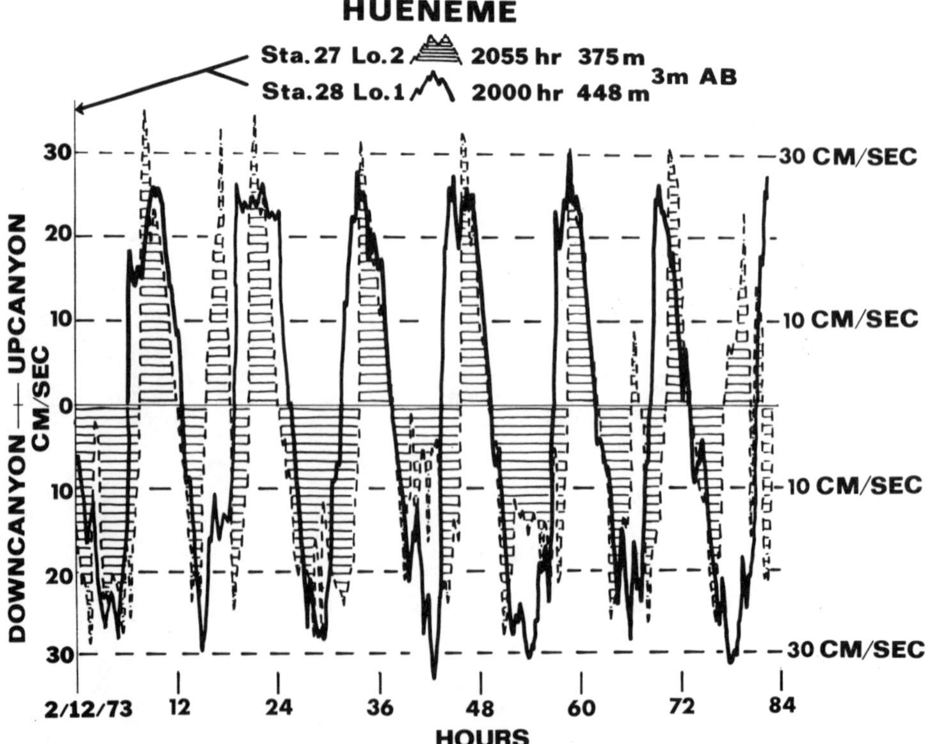

FIG. 51—Superposition of contemporaneous time-velocity curves from 375 and 448 m in Hueneme Canyon. The best fit was obtained by a shift to later time at the shallow station, indicating upcanyon advance of internal waves.

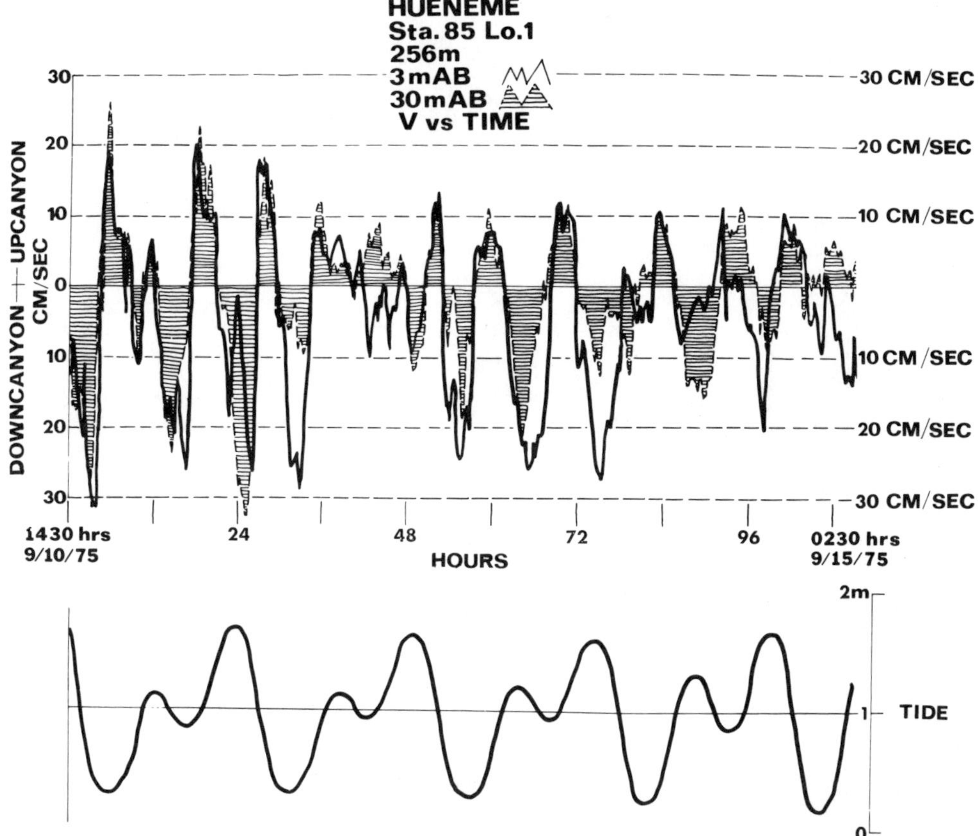

FIG. 52—Overlay of time-velocity series curves at 3 and 30 m taken at a depth of 256 m in Hueneme Canyon. Curves in the first part agree both in speed and times of direction change, but later in the record there is somewhat progressive disagreement.

they match fairly well by moving the curve of the deeper station so that the times are 3 hours and 45 minutes later at the deeper than at the shallow station (Fig. 54). Thus the indications are that internal waves were moving down Santa Cruz Canyon. The net flow at the deeper station also is downcanyon, and the fastest velocities, 29 cm/sec, are in a downcanyon direction. However, the fastest velocity for the shoaler station is 36 cm/sec in an upcanyon direction.

In 1975 we obtained our records during relatively calm conditions with gentle northwest winds, but the currents at the shoaler stations are somewhat faster than during the 1973 recordings. In the 1975 records the alternations at the 208-m depth are of shorter period than tidal, but they show some slight relationship to the diurnal tide (Fig. 55B). The net flow is downcanyon.

In 1975 at the deeper station the currents are again slower than at the shallow stations with a maximum of 30 cm/sec and the net flow is again downcanyon. Comparing the two 1975 curves we find the best fit is for the deeper station to be 3 hours and 53 minutes later than for the shallow station (Fig. 54B). This also indicates downcanyon advance of the internal waves. The fit is not as good as for the

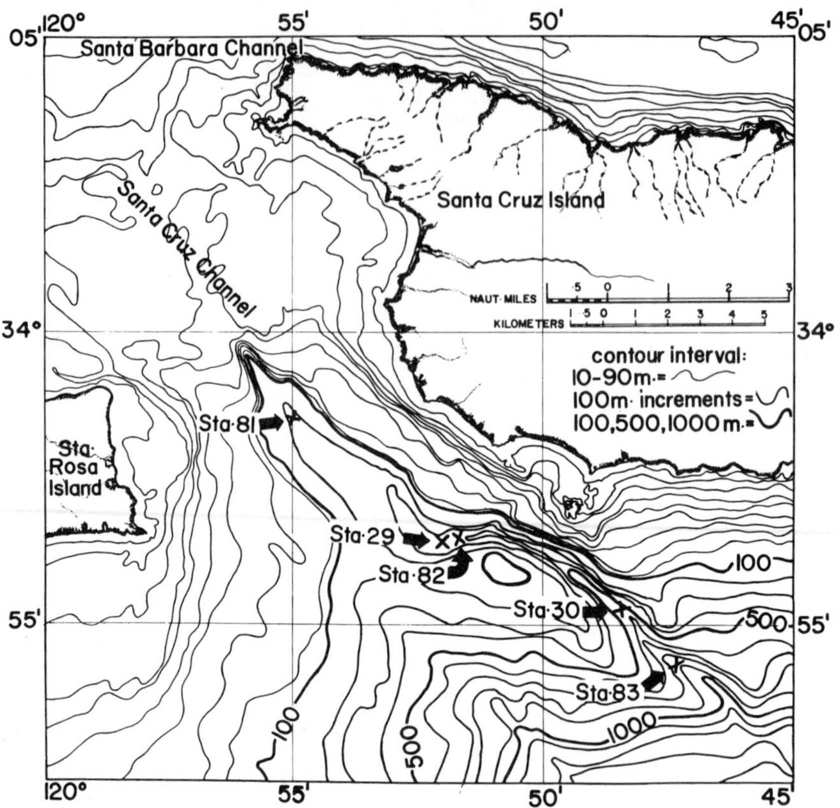

FIG. 53—Contour chart of Santa Cruz Canyon showing axial bends and location of current-meter stations. Notice relation of canyon head to strait between the islands. For relation to Santa Barbara Channel see Figure 46B.

two earlier stations. The advance of internal waves in that direction is supported by direction changes in the record for the intermediate station (with no velocity). It is possible to find a rough fit from comparing the direction changes, and this also indicates downcanyon advance of the internal waves.

Carmel Canyon

Unlike other submarine canyons along the California coast, Carmel Canyon extends into a dendritic bay, and the main head forms in an indentation directly off the mouth of San Jose Creek at the south end of Carmel Bay. There are a group of tributaries entering this canyon head (Fig. 56), and the canyon is narrow and V-shaped with rock walls, mostly granite, and in general it has a sandy or rocky floor. Like many other California canyons the head is so close to the shore that a stone can be thrown into it from the beach. A few comparative surveys made by Shepard showed that the head of Carmel Canyon starts to fill rapidly with sediment and is then cleaned out as in other canyons. The axis at the canyon head trends west, but outside Carmel Bay it bends to the north-northwest and enters Monterey Canyon at a depth of about 2,000 m (Fig. 57).

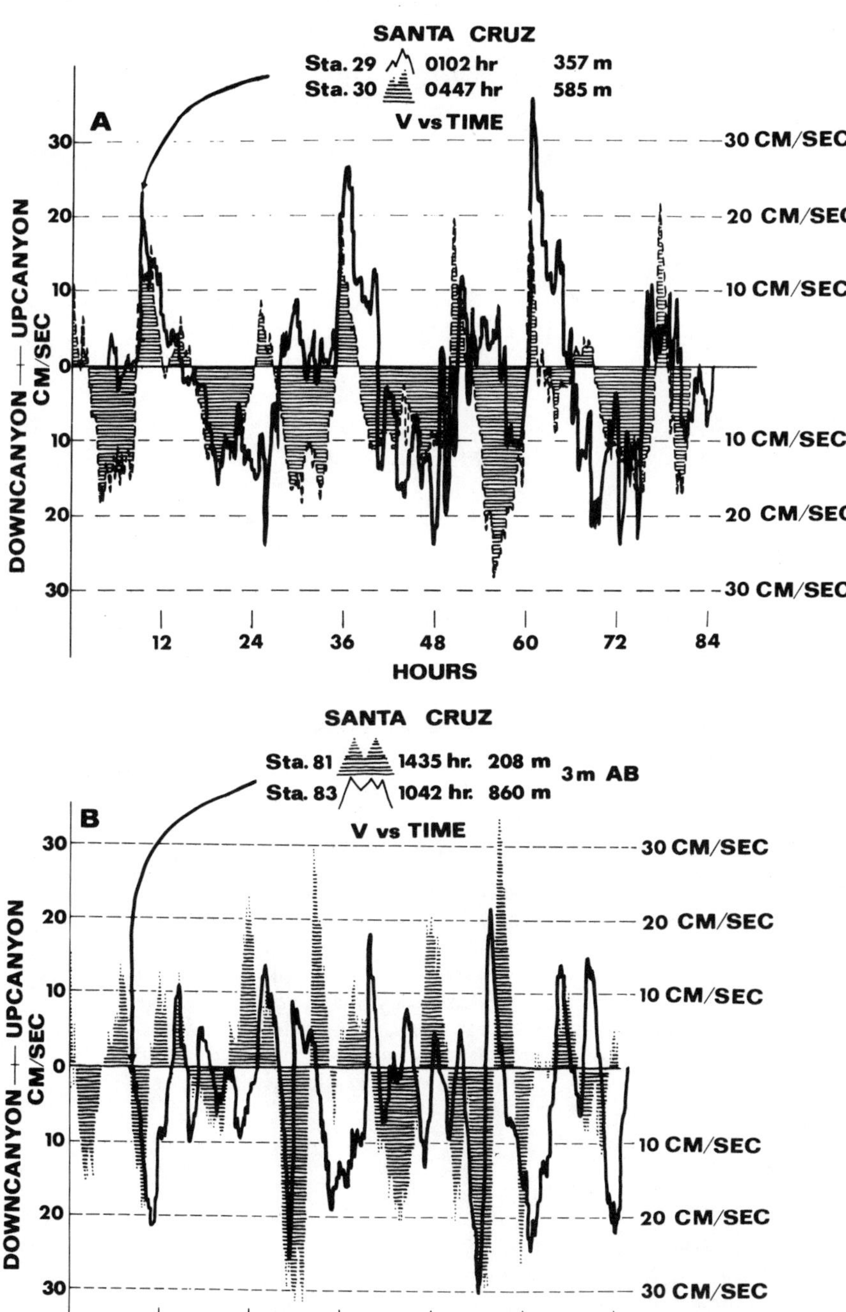

FIG. 54—Superposition of adjacent stations taken contemporaneously for 357 and 585 m (**A**) and for 208 and 860 m (**B**). In both cases the best fit is for a time shift to a later time for deeper station and hence for downcanyon advance of internal waves. Fit is not very good, but the relation to the tide helps confirm the interpretation. Here the introduction of water from Santa Barbara Channel through the strait between the islands is believed to be the reason for advance of the internal waves downcanyon.

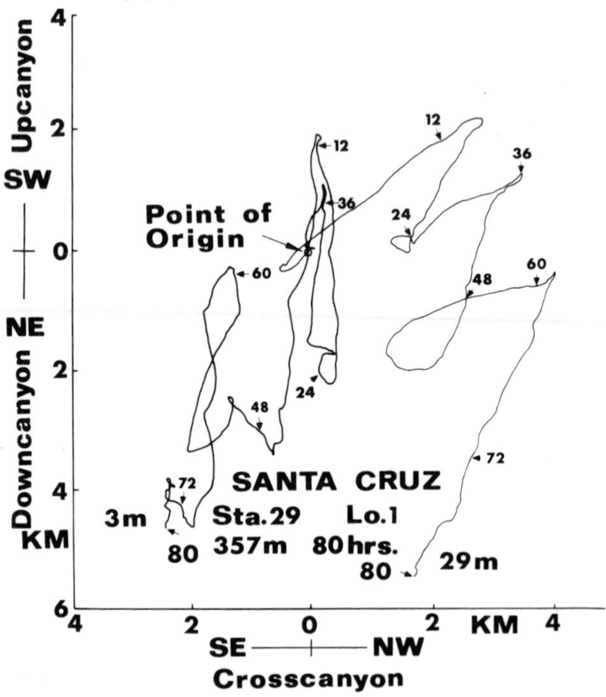

FIG. 55—Progressive-vector diagram of currents at 357-m depth in Santa Cruz Canyon at 3 and 29 m above bottom shows possible influence of southerly wind during first part of record and of stronger northwest wind during last part. Direction of flows changed shortly after hour 36 on both meters.

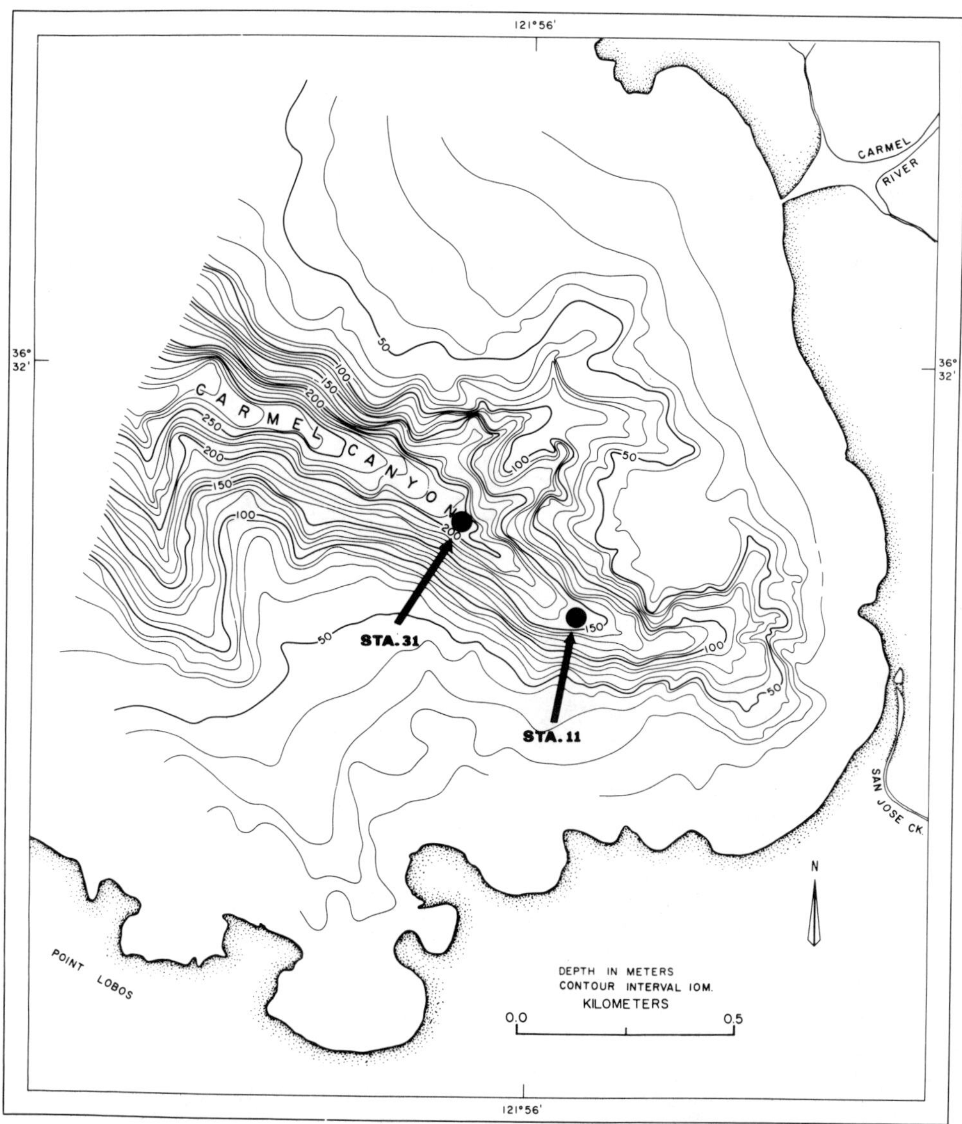

FIG. 56—Large-scale contour chart of the head of Carmel Canyon. Principal branch extends into the cove at the mouth of San Jose Creek rather than into the mouth of Carmel River. Two current-meter stations are shown (contours from a 1934 survey by Shepard and from U.S. Coast and Geodetic Survey). See also Figure 57 for seaward continuation of the canyon.

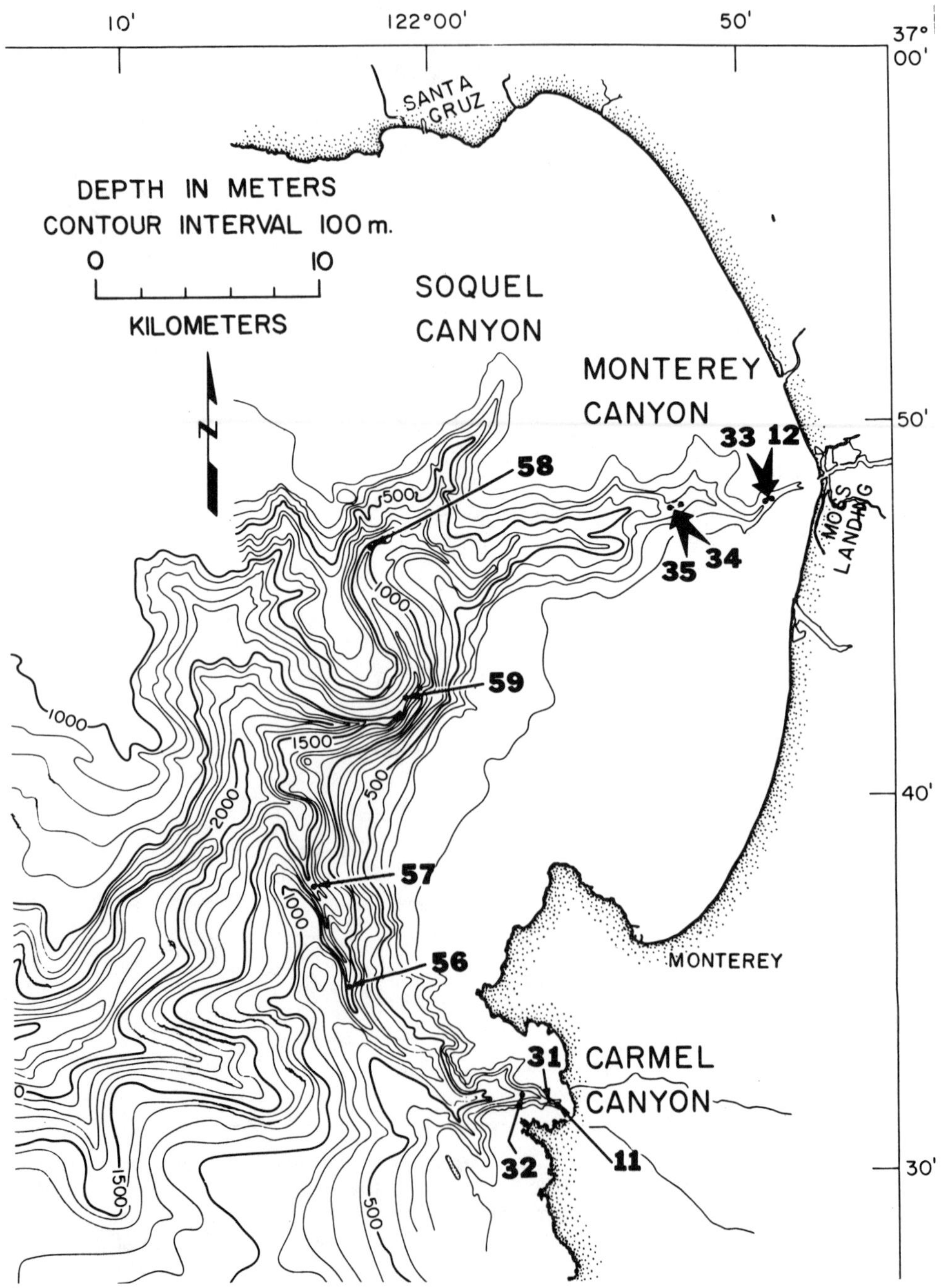

FIG. 57—Contours of Monterey and Carmel Canyons; note the meandering course of Monterey Canyon. The current-meter stations of both canyons are indicated.

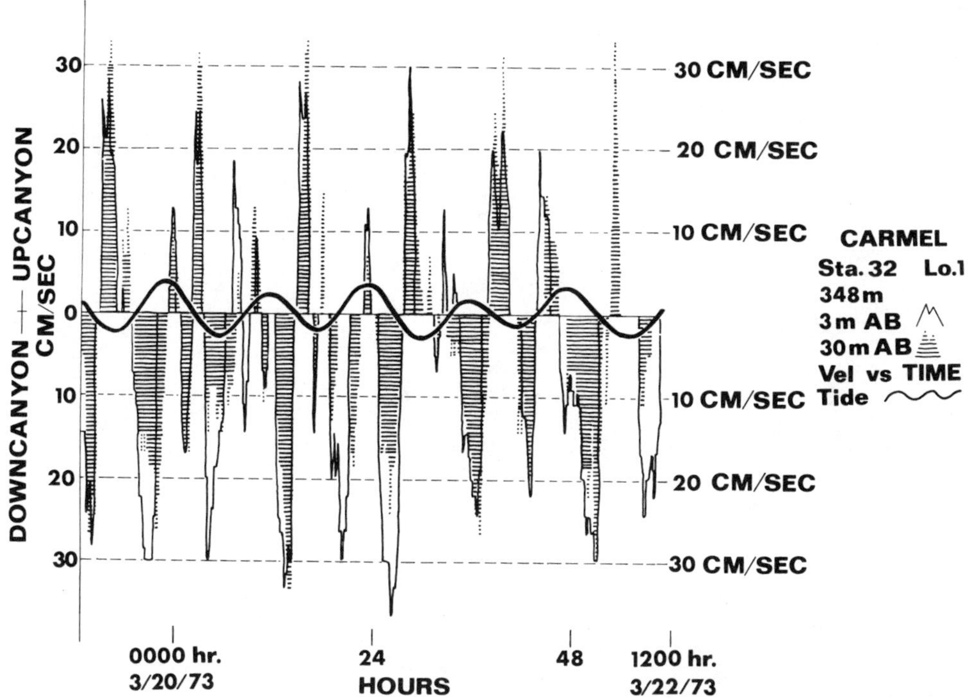

FIG. 58—Examples of remarkably similar time-velocity curves at 3 and 30 m above bottom at the 348-m station in Carmel Canyon. Agreement is good both in times of current direction change and in speed of currents. This canyon is unique in having such good agreement of records from overlying current meters (see also Fig. 18).

We obtained nine current-meter records from the canyon, thanks to cooperation of the U.S. Navy Postgraduate School. Stations were located at 156, 205, 348, 1070, and 1445 m, thus covering much of the length of the canyon. All but the shallowest station had two current meters. The three deepest stations had meters at 3 and 30 m above bottom and the station at an axial depth of 205 m had current meters at 3 and 19 m.

The most striking feature in the records at Carmel Canyon is the remarkable similarity between currents at 3 and 30 m (or 3 and 19 m) at the four stations (Figs. 58, 18). The velocity at the two depths was generally very similar and the time of change from up- and downcanyon flows was almost simultaneous. In all our investigations at these two rather well-separated levels we have never found such close agreement. A strange thing about these similarities is that the polar plots for 3 m above bottom at the 348-m station show the upcanyon flow is by no means a reciprocal of the downcanyon (Fig. 59A). This is not true of the 30-m above bottom record (Fig. 59B). However, in both cases the currents changed direction at about the same time, and the rate of flow was very similar. The difference may be a variation in up- and downcanyon direction caused by a curve of a narrow canyon gorge which may exist entirely below the 30-m level.

The crosscanyon plot at 348 m shows an interesting relation to the tide. All but one of the relatively strong north crosscanyon flows occurred during an ebb-tide period (Fig. 60).

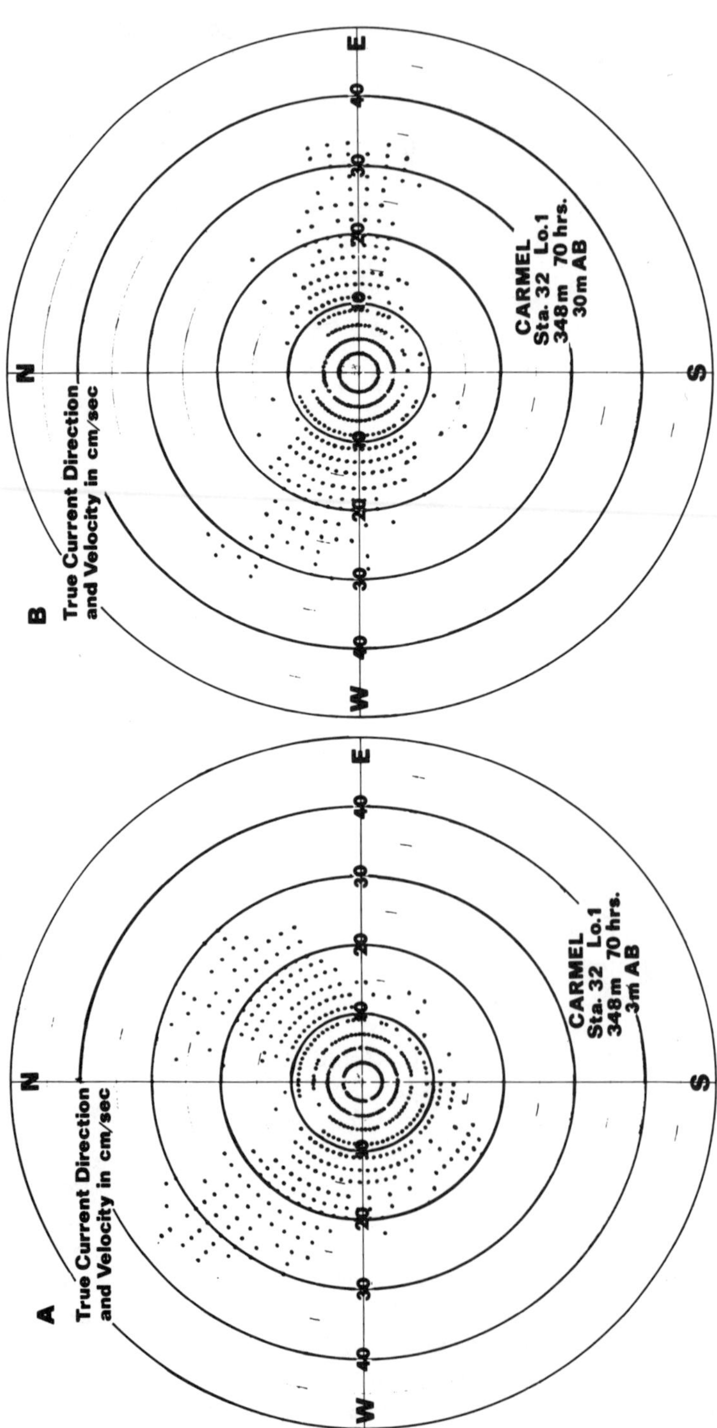

FIG. 59—Polar plots of records from Carmel Canyon at depth of 348 m. At 3 m above bottom the up- and downcanyon directions are at right angles, whereas at 30 m they are nearly opposite or 180° apart. Possibly the narrow gorge at the bottom had a sharp curve that deflected up- and downcanyon currents differently and the current meters at 30 m above bottom may have been above the effect of this incised gorge.

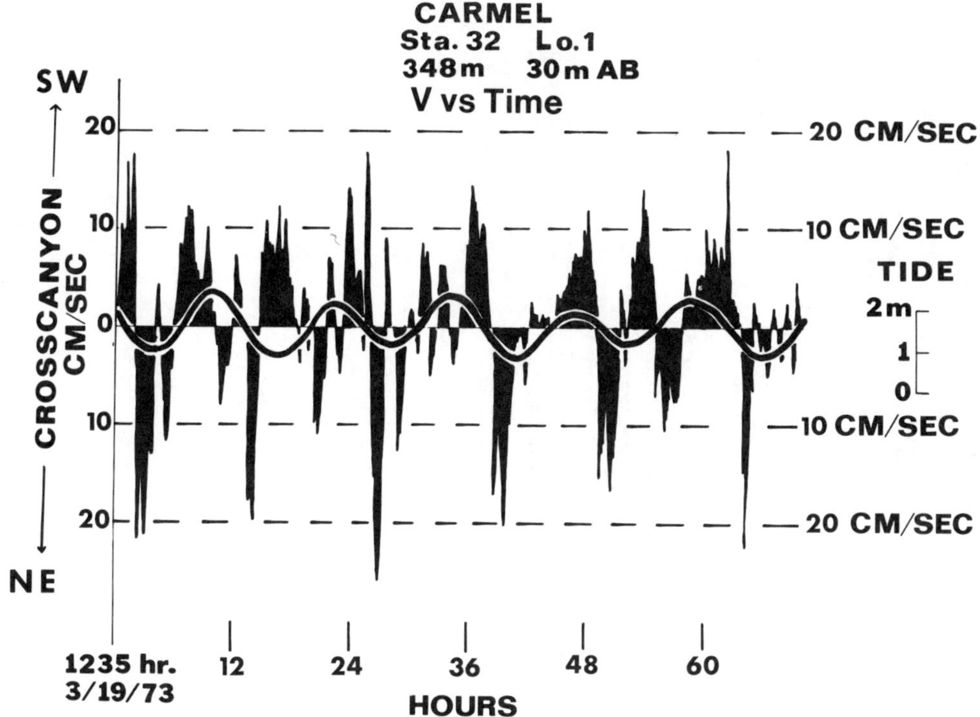

FIG. 60—Figure illustrates the relation of the crosscanyon maximum flows to ebb-tide periods in the record from 348-m depth in Carmel Canyon, possibly a coincidental relation.

Carmel Canyon records are also unique in having such good comparisons with neighboring stations. The offset times always suggest upcanyon progress of internal waves. This was not surprising in comparing stations at 205 and 348 m because they were only separated by 1.3 km, but stations at 1,070 and 1,445 m have a separation of 6.3 km, and these compare just about as well (Fig. 61). Because the records at 3 and 30 m above bottom are so comparable at all these stations, it also is possible (timewise) to compare currents at different heights above the bottom with those at neighboring stations. These comparisons agree nicely.

The average lengths of up- and downcanyon reversal cycles at 3 m above bottom show the following: at 156 m, 3.6 hours; at 205 m, 4.1 hours; at 348 m, 5.1 hours; at 1,070 m, 10.2 hours; and at 1,445 m, 11.7 hours. The two deepest stations have a clear relationship to the tidal curves (Fig. 61B), and the record at 348 m (Fig. 58) shows some tidal effect. Thus the tidal period becomes dominant at a little greater depth in Carmel Canyon than in most of the other California canyons with comparable tidal ranges.

Storm conditions with strong onshore winds seem to produce turbidity currents in La Jolla and Scripps Canyons, but in Carmel Canyon when we launched (with considerable difficulty) current meters at 1,070 and 1,445 m during winds that reached about 30 knots (55 km/hour), there was no such current at these stations. Figure 61B does not show any downcanyon speed exceeding 30 cm/sec. Perhaps the

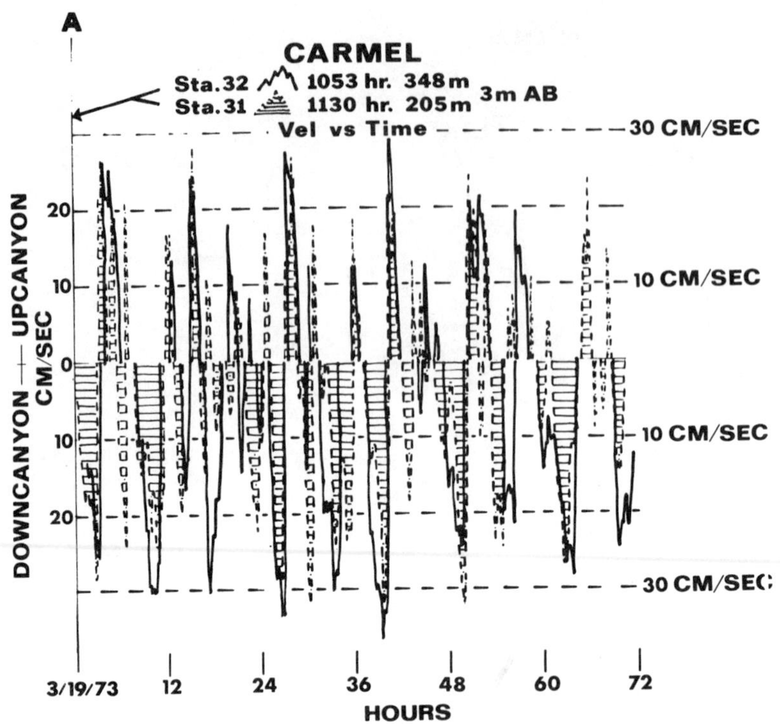

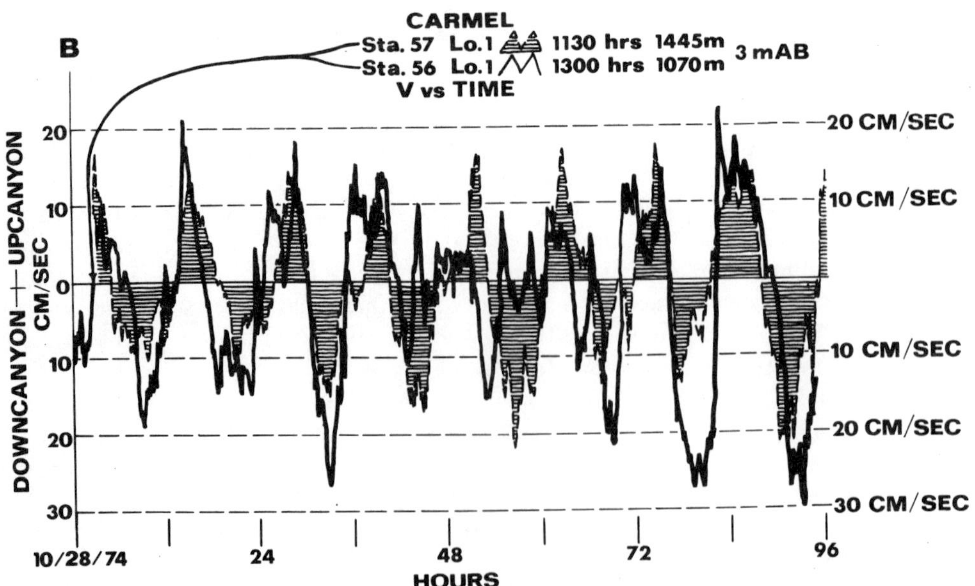

FIG. 61—**A**, Superposition of time-velocity curves from 205 and 348-m depths in Carmel Canyon. Best fit time shift so that shoaler station is 37 minutes later. This is one of the best agreements in any of our records and shows the internal waves were moving up Carmel Canyon. **B**, Superposition of time-velocity curves from 1,070 and 1,445-m depths in Carmel Canyon showing another good fit indicating upcanyon advance of internal waves in the deeper part of this canyon.

storm was of too short a duration or a turbidity current in the canyon head may not have extended this far down the canyon.

Monterey Canyon

Monterey is the largest submarine canyon on the California coast, although like most of the other California canyons it is not now located off a river (Fig. 57). Like the others, it extends very close to the shore. One canyon head is so close to the Moss Landing pier that it caused structural damage to the pier when slides or turbidity currents occurred. Another head lies directly off the entrance to Moss Lagoon. It is claimed that there was a river from the Great Valley of California entering at this point in the Pleistocene (Jenkins, 1973). The submarine canyon has a meandering course and is cut partly through granite, but the walls are largely sedimentary (Martin and Emery, 1967; Greene, 1970, 1977). Even the fan valley outside includes a meander (Shepard, 1966). A number of tributaries enter Monterey Canyon including Carmel and Soquel Canyons. In some cross sections, the canyon has walls that are as much as 2,000 m high. At least one basin depression occurs along the course of the inner canyon. This was discovered in 1970 from a deep-diving vehicle by Gary Greene and Edward Clifton of the U.S. Geological Survey (Greene, 1977, p. 140).

We investigated currents in Monterey Canyon on three occasions obtaining nine records. Three earlier records, obtained by the U.S. Navy Postgraduate School, were puzzling to us because they failed to show any concentration of current flow along the canyon axis. The Navy records were all made in shallow water, but in our records, which included depths up to 1,445 m,[3] we found five which also had virtually no flow direction related to the canyon axes, and four which lack much concentration in up- and downcanyon directions (Fig. 62). Section G in Figure 62 shows some of the fastest currents to the southwest and lacks slower currents in the same direction as do two records in Hueneme Canyon. This lack of directional control appears to be a characteristic of Monterey Canyon at least out to 1,445 m, our deepest station. An explanation for this lack is not yet evident. As mentioned previously, the right angle between up- and downcanyon currents in Carmel Canyon appears to be related to a curve in the canyon at the point where the currents were measured. Possibly the same explanation applies to much or all of Monterey Canyon, which certainly appears to have a decidedly curving axis. Also there may be many more tributaries than indicated in the available charts. Currents moving in and out of these small tributaries could change the current directions. The rather detailed soundings which we took along a part of Monterey Canyon failed to clarify the situation. However, future descents in deep-diving vehicles may show that the topography is extremely complicated, thus accounting for the contrast in currents between this canyon and those of other canyons that we have investigated.

The progressive-vector diagrams of the Monterey data show various patterns and on the whole are more irregular than in most of our other canyon records (Fig. 63). Unlike most California stations, the net flow is up in 6 out of the 9 records, including one record obtained 30 m above the bottom with the others at 3 m.

In one respect the Monterey records can be compared favorably to those of other

[3]The duplication of a station depth in Carmel Canyon is such a strange coincidence as to suggest an error, but according to the chart it is about the right order of magnitude.

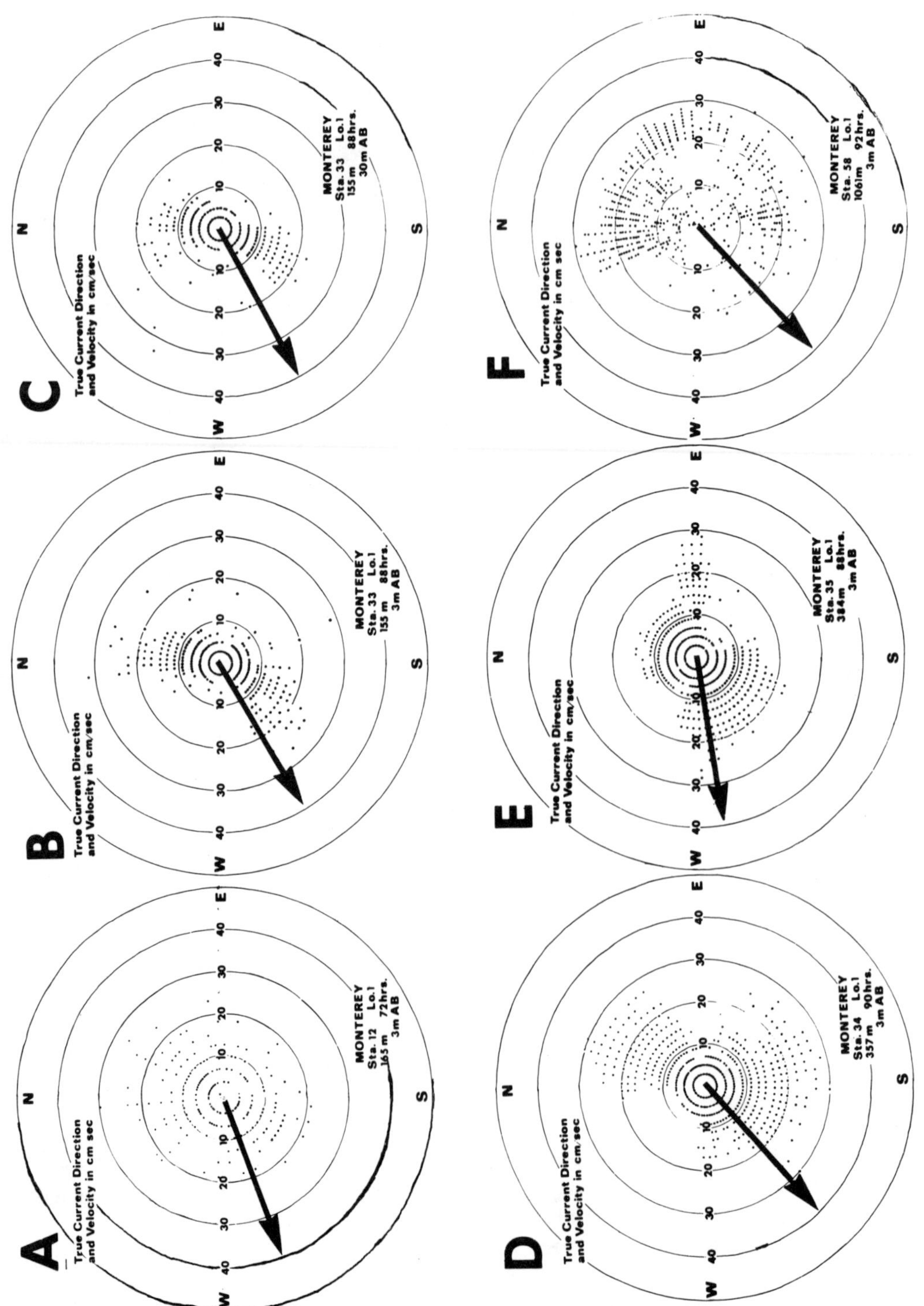
A
True Current Direction and Velocity in cm sec
MONTEREY Sta. 12 Lo.1 165m 72hrs. 3m AB
B
True Current Direction and Velocity in cm/sec
MONTEREY Sta. 33 Lo.1 155 m 88hrs. 3m AB
C
True Current Direction and Velocity in cm/sec
MONTEREY Sta. 33 Lo.1 155 m 88hrs. 30m AB
D
True Current Direction and Velocity in cm/sec
MONTEREY Sta. 34 Lo.1 357m 90hrs. 3m AB
E
True Current Direction and Velocity in cm/sec
MONTEREY Sta. 35 Lo.1 384m 88hrs. 3m AB
F
True Current Direction and Velocity in cm sec
MONTEREY Sta. 58 Lo.1 1061m 92hrs. 3m AB
N
S
E
W

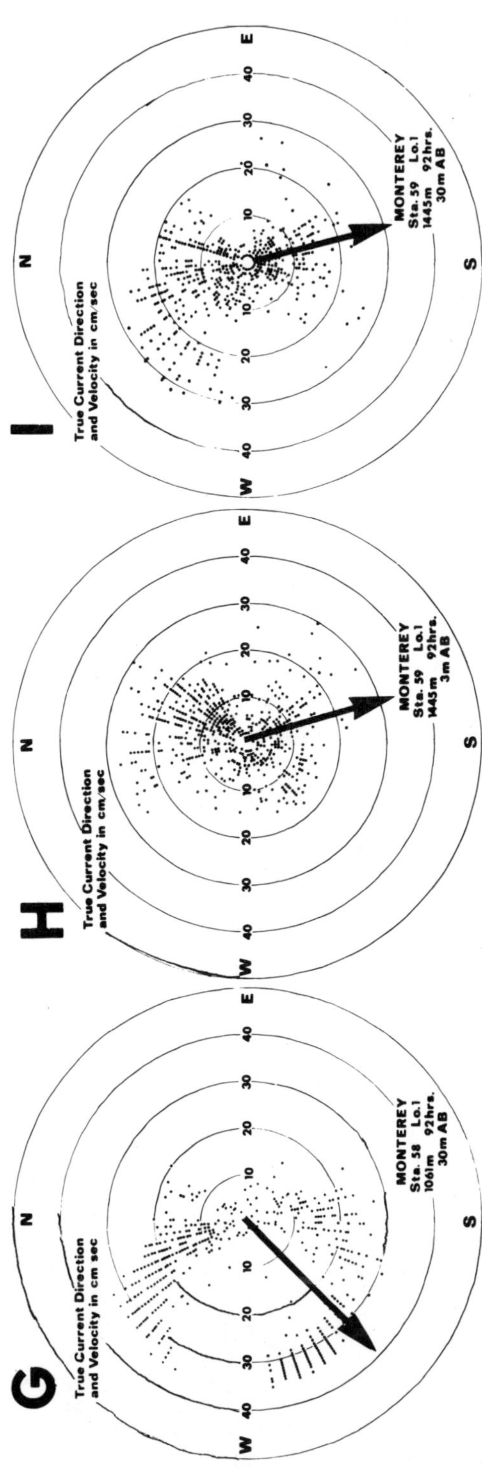

FIG. 62—Polar plots of the records from Monterey Canyon showing the almost complete failure of currents to conform to the axis of the canyon. Downcanyon direction as determined from soundings is indicated by an arrow. Some of the faster flows as in **F** and **G** have no intermediate or slow currents in the same direction. Compare the crosscanyon flows in Hueneme Canyon (Fig. 47) where the same is true.

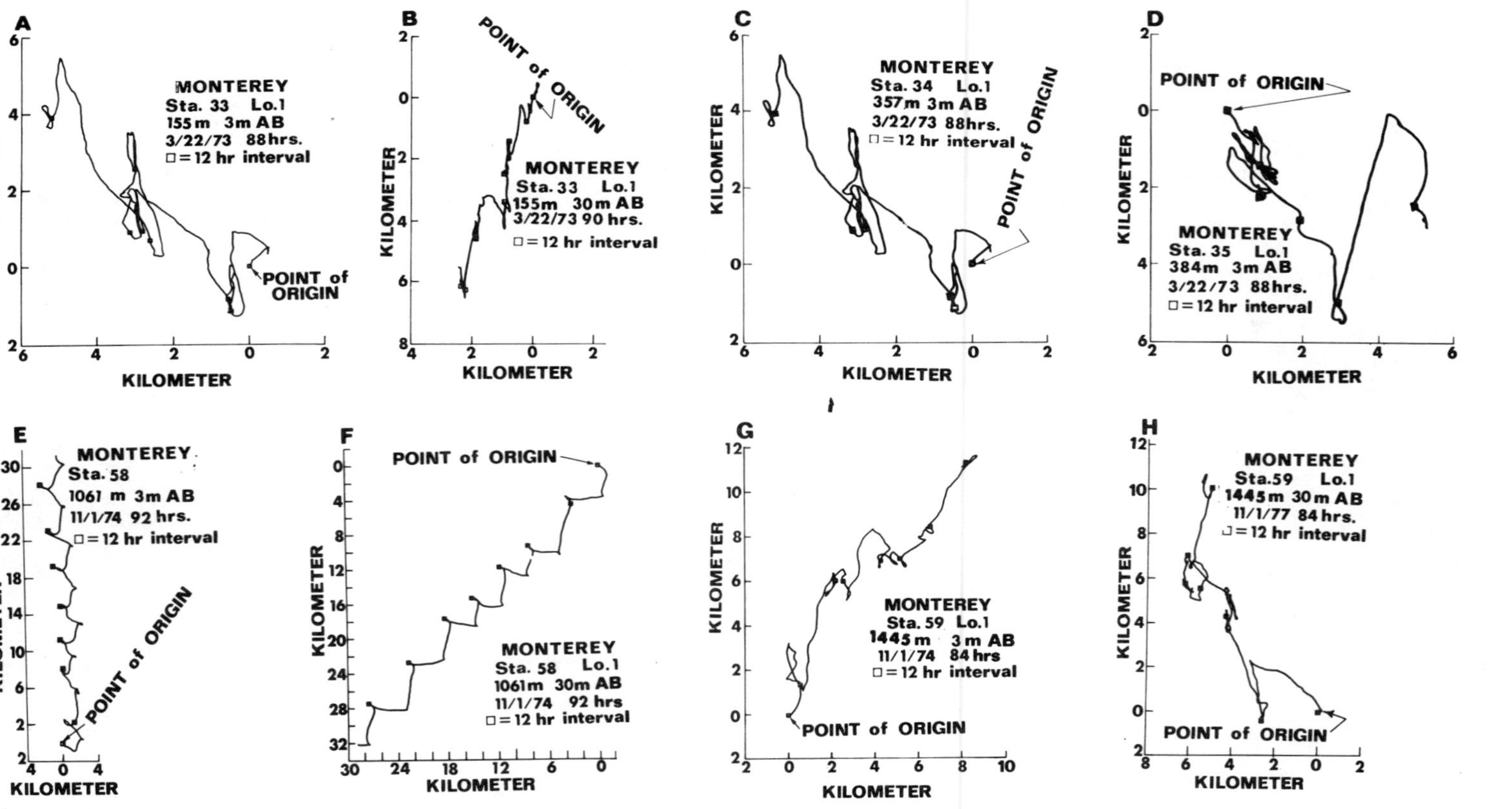

FIG. 63—Progressive vector diagram of currents in Monterey Canyon, showing the great variation in flow pattern. Five out of the 8 show net flow upcanyon, unique for canyons in general and especially for California canyons.

canyons. Overlaying the 3- and 30-m records at the deepest station (1,445 m), we find that they are quite similar (Fig. 64B), much as they are for the Carmel stations. However, at 155 m the 3-m net flow is upcanyon, and at 30 m it is downcanyon (Fig. 63A,B).

In still another respect Monterey Canyon is exceptional. The two stations at 155 m show (respectively) average cycle lengths of 4.4 hours for 30 m above bottom and 7.2 hours for 3 m above bottom. The average at 357 m is 8.8 hours; at 384 it is 8 hours; at 1,061 m both the 3-m and 30-m above bottom averaged 6.5 hours; and at 1,445 m the 3-m level had an 8.7-hour average and the 30-m had a 10-hour average. Thus there is no detectable increase in the length of the cycles with depths greater than 155 m. On the other hand many of the deeper stations have the tidal period represented in some parts of the alternation cycles (Fig. 65).

Finally the Monterey Canyon stations show internal wave patterns indicating upcanyon advance of internal waves between stations at 155 and 368 m and between 368 and 384 m (Fig. 66) but downcanyon advance between stations at 1,061 and 1,445 m (Fig. 66).

San Clemente Rift Valley

The fault scarp along the landward (northeast) slope of San Clemente Island off southern California becomes a rift valley to the southeast, and extends to the San Clemente Basin (Fig. 67). The floor of this V-shaped valley slopes continuously to the southeast, extending in depth from approximately 1,000 to 2,000 m. A dive into this valley at about 1,200 m was made by D. G. Moore in *Deepstar 4000* (Moore, 1969, p. 55). Although the currents were negligible while Moore was observing the bottom, he found clear evidence that currents had been flowing upvalley, producing ripple marks with steep sides to the northwest and other indications of the same direction of flow.

Since this rift valley is located about 90 km west of the California coast and extends essentially parallel with the coast, it forms a marked contrast to nearby submarine canyons. Thus, it was a particularly interesting place to make a bottom current survey. Accordingly, with the use of *Velero IV*, on a trip with John Warme as chief scientist, we placed current meters in the valley axis at depths of 1,372, 1,646, and 1,756 m, all instruments at 3 m above the bottom. The middle station was maintained for only 4 days, but the other two were in operation for a month.

All three records showed that the currents have an alternation of up- and downvalley flows and that the flow cycles are clearly related to either the semidiurnal or diurnal tide (Fig. 68). However, the diurnal period is only indicated in the time-velocity curves at the deepest station, and even there, minor up- and downvalley pulses have a semidiurnal tidal period. Another example of this diurnal tide relation occurred in one of the shallow stations of Santa Cruz Canyon as has been mentioned. Although all of the rift valley time-velocity plots show relations to the tidal curve, direction reversals are, as usual, at somewhat more irregular intervals than the tides. The average period of alternation at the 1,372-m station is about 11 hours; at 1,646 m, about 12 hours; and at 1,756 m about 19 hours (Fig. 68). Thus the period increases with depth although it is dominantly related to the tide at all stations. The 19-hour period is the result of a diurnal (24 hours, 52 minutes) period with some semidiurnal pulses interspersed.

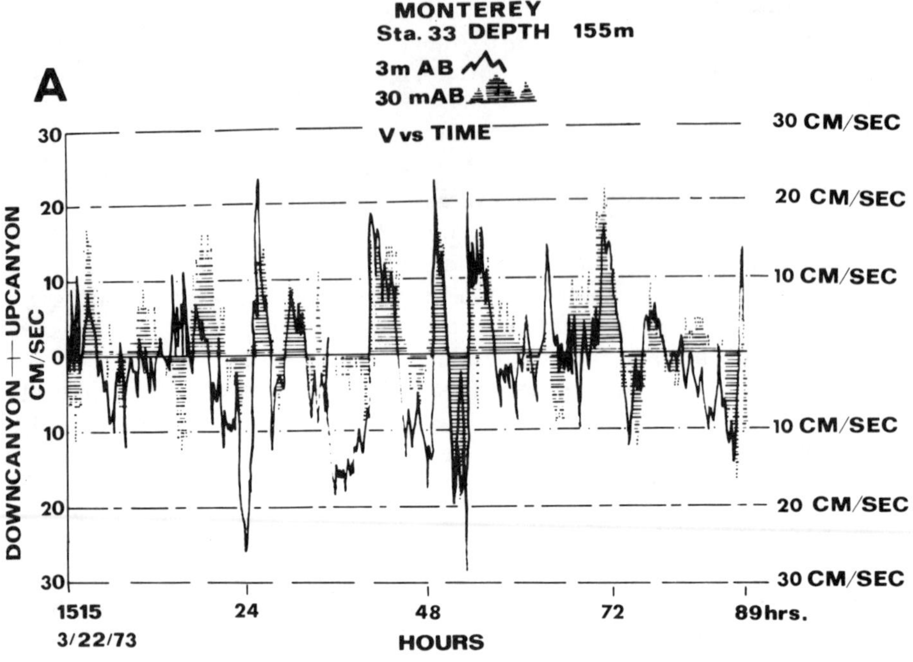

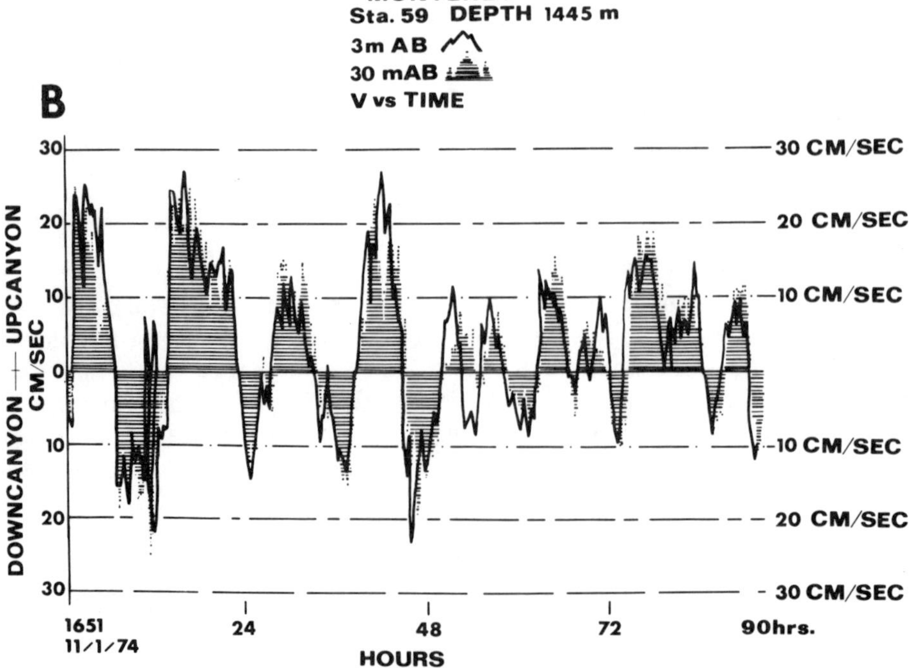

FIG. 64—Overlay of time-velocity curves at 3 and 30 m in two stations in Monterey Canyon, **A**, at 155 m, shows general agreement of times of up- and downcanyon but considerable difference in current speeds at the two levels. **B**, at 1,445 m, has excellent agreement in both time and speed of up- and downcanyon currents at the two levels with few exceptions.

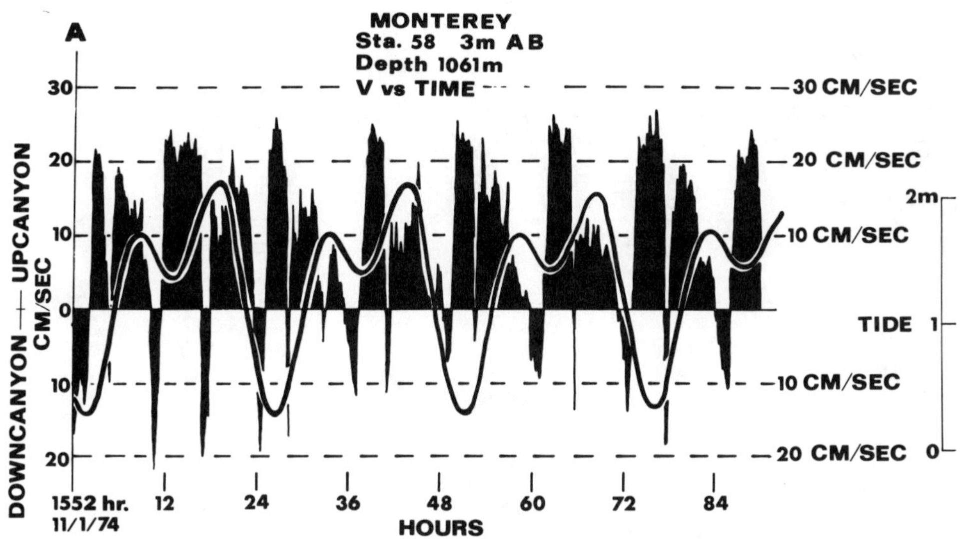

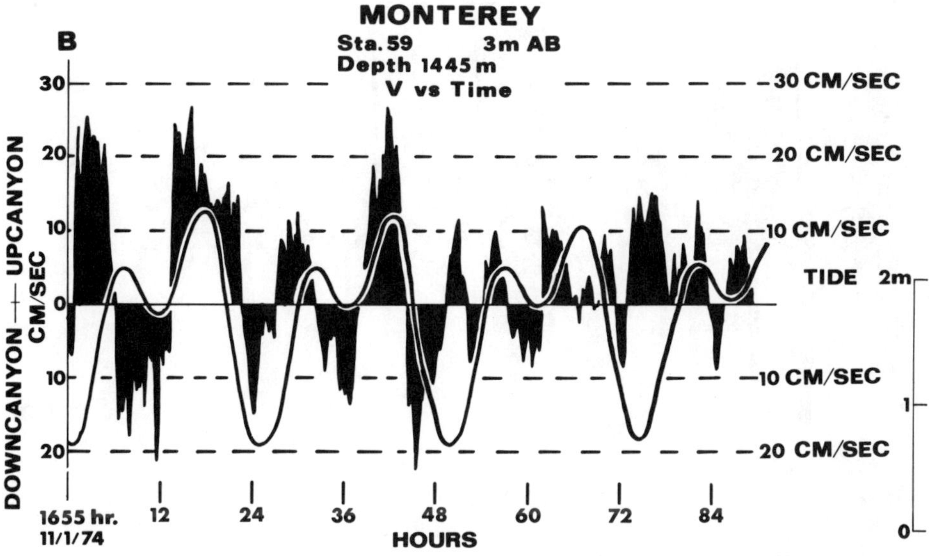

FIG. 65—Figure shows rather poor relation of the currents to the tides at the two deepest Monterey Canyon stations. **A**, at 1,061 m, shows a rough tidal relation of upcanyon major flows but to only a few downcanyon flows. **B**, at 1,445 m, shows the possible relations to the tides with some diurnal influence for upcanyon flows and some semidiurnal relations to most downcanyon flows.

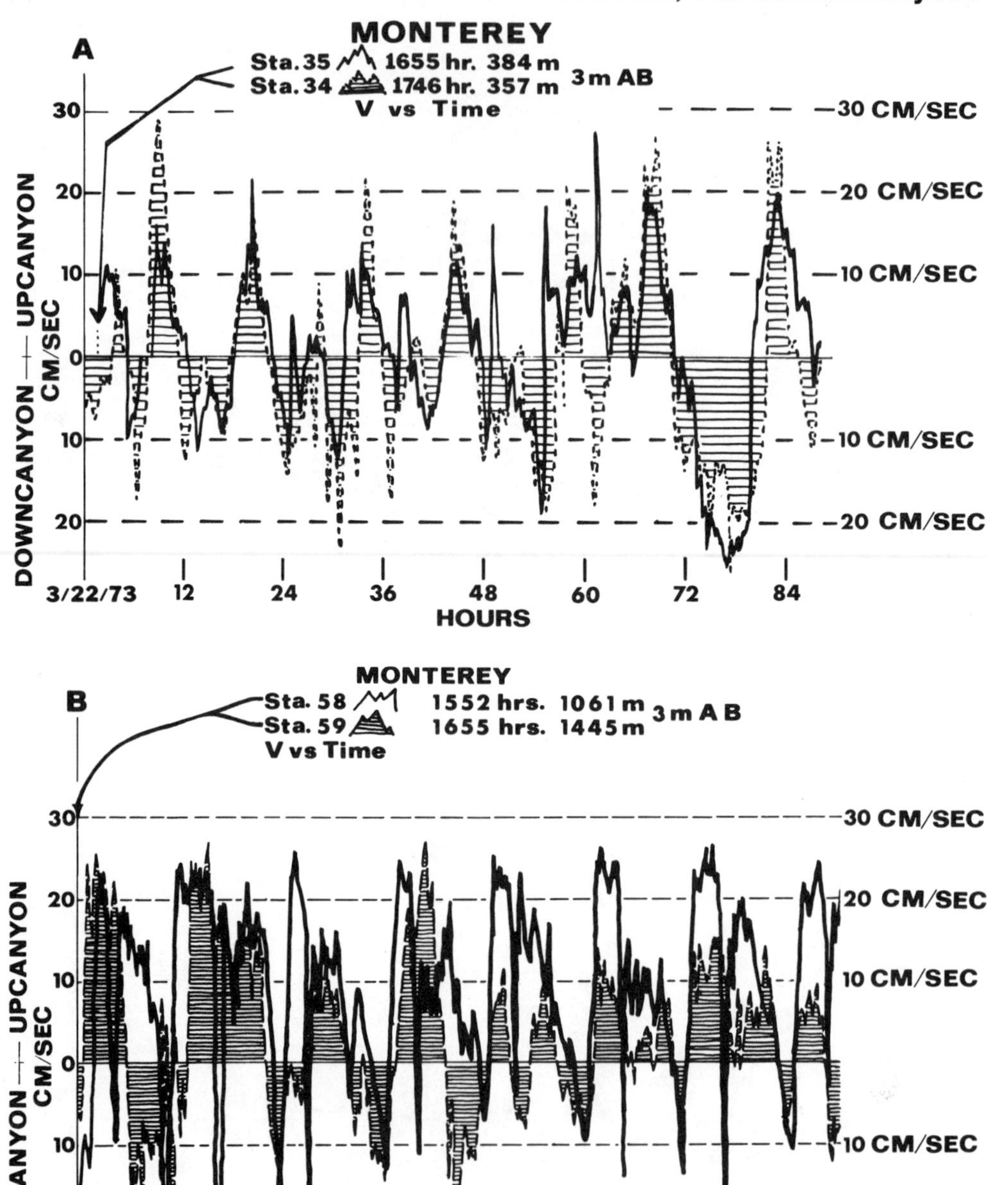

FIG. 66—**A**, superposition of time-velocity curve for Monterey Canyon at 357 and 384 m depths. The matching of the curves at these two closely spaced stations is as good as expected but some glaring exceptions exist as at hour 62 where there was a strong upcanyon flow at the deeper station and moderate downcanyon flow at the shallow station; otherwise both speeds and times match well. Best fit is 51 minutes later at the shallow station indicating upcanyon advance of internal waves. **B**, superposition of time-velocity curves of Monterey Canyon stations at 1,061- and 1,445-m depths. These stations were separated by about 8 km and the

FIG. 66—(continued)

agreement is not as good as in the shallower and closer spaced stations. The time shift indicated later arrival at the deeper station, unlike the time difference for the shallower pair. Monterey is the only canyon where we found evidence that internal waves were advancing in opposite directions, however, the measurements for the two pairs were made at a different time.

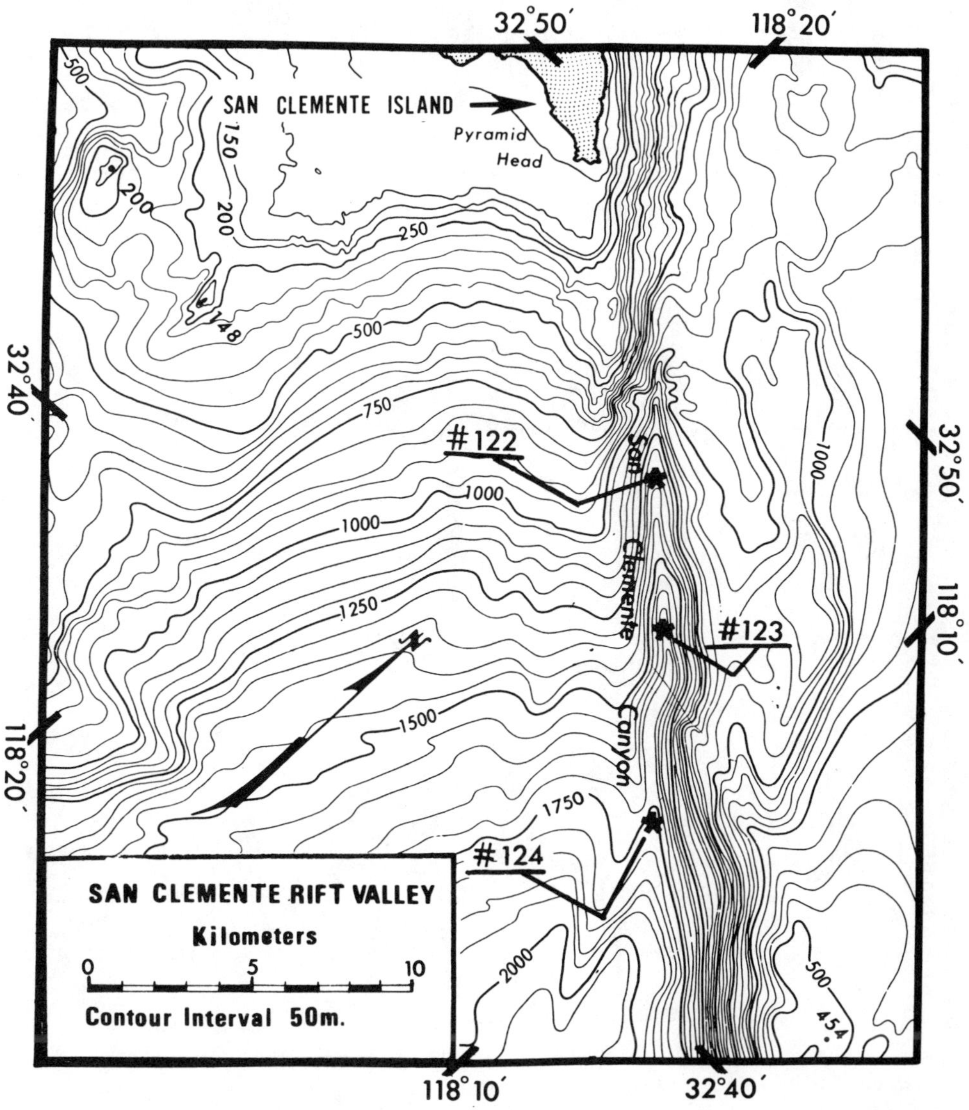

FIG. 67—Contour chart of San Clemente Rift Valley 80 km off San Diego. Current-meter station locations are included (from contour chart by NOAA). Station 124, where diurnal tide effects were indicated in the time-velocity curve, is beyond the deeply incised part of the fault valley (Fig. 68).

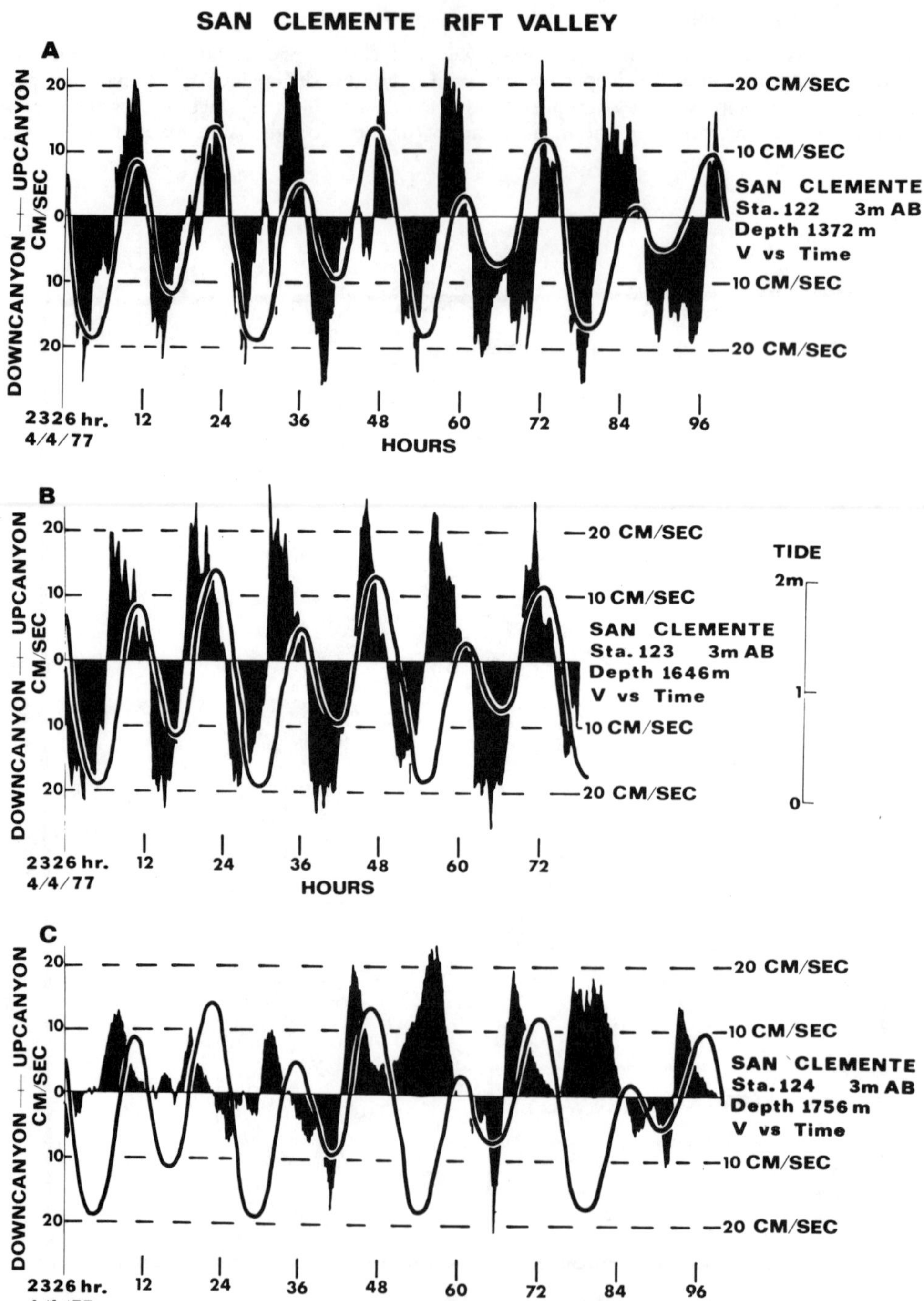

FIG. 68—Time-velocity curves of contemporaneous portions of the records in San Clemente Rift Valley showing the clear tidal relations. The maximum flows show a progressive relation to the tide curves indicating an upvalley advance of internal waves.

Comparing the time-velocity curves for the three stations, we found a progressive shift making the times successively later at each upvalley station. This is particularly well demonstrated for the month-long records at the shallow and deep stations (Fig. 69). The curves can be matched along the entire length with an offset of about 2 hours and 53 minutes. This advance over a distance of 12 km is somewhat faster than we usually find for internal waves. The best match for the shallow and intermediate stations is only based on four days, but again it indicates upvalley advance as does the best match for the intermediate and deep stations.

In view of the earlier indications from a deep-diving vehicle, the net flows at the three stations are of more than passing interest. We found that the net is downvalley for the shallow and intermediate stations, and upvalley for the deeper station (Fig. 70).

The polar plots are all indicative of flow almost directly up- and downvalley as would be expected for this narrow, straight rift valley. Similarly, crossvalley plots indicate very small transverse currents.

The velocity of the up- and downcanyon flows is about the same as for those at similar depths in California canyons. Speeds are distinctly slower at the deepest station, as we found in our equally deep stations in Monterey and Carmel Canyons.

These data indicate that currents in this strike-slip fault valley, located well out from the coast and extending parallel with it, have a close resemblance to those of the coastal submarine canyons.

Fraser Seavalley, British Columbia

In the marine valleys discussed previously, we dealt with a series of submarine canyons and one rift valley. The next valley to consider is off the Fraser delta at the mouth of the Fraser River. It is a seavalley of slight depth of the type found in the slopes off most deltas where deposition has built across a continental shelf (discussed in Shepard, 1955, 1973; Mathews and Shepard, 1962; Shepard and Milliman, 1978). This valley creases the slope directly outside the mouth of the Fraser River and can be traced down to a depth of about 200 m (Fig. 71A). A more detailed chart of the head of this valley was made by the Public Works of Canada (Fig. 71B). Both charts show a series of 3 valleys, each 20 to 40 m deep with an average width of about 500 m. The northernmost of the three extends landward as far as the jetty at Sand Head and includes several smaller tributaries. Other valleys appear further north but are not as well sounded. In earlier investigations (Mathews and Shepard, 1962) a group of mounds discovered on the lower deltaic slopes were interpreted as slump deposits formed by slides on the advancing delta foreset slope. This was confirmed by Tiffin et al (1971) from seismic profiles.

We measured currents in the principal valley during April and May of 1976 and in early June of 1977, and both occasions included times of large tidal range (4 m). In the first period the melting snows in the mountains produced a high flow of water down the Fraser, whereas during the second period there was little winter snow so that the runoff was relatively low for that time of year. In both operations we placed current meters at three depths in the valley, but in the first deployment several of our current meters did not operate completely, whereas in the second, they all yielded good records. At the end of the first period we lost one current meter during the highest spring flow of the Fraser River. On this occasion the current meter was

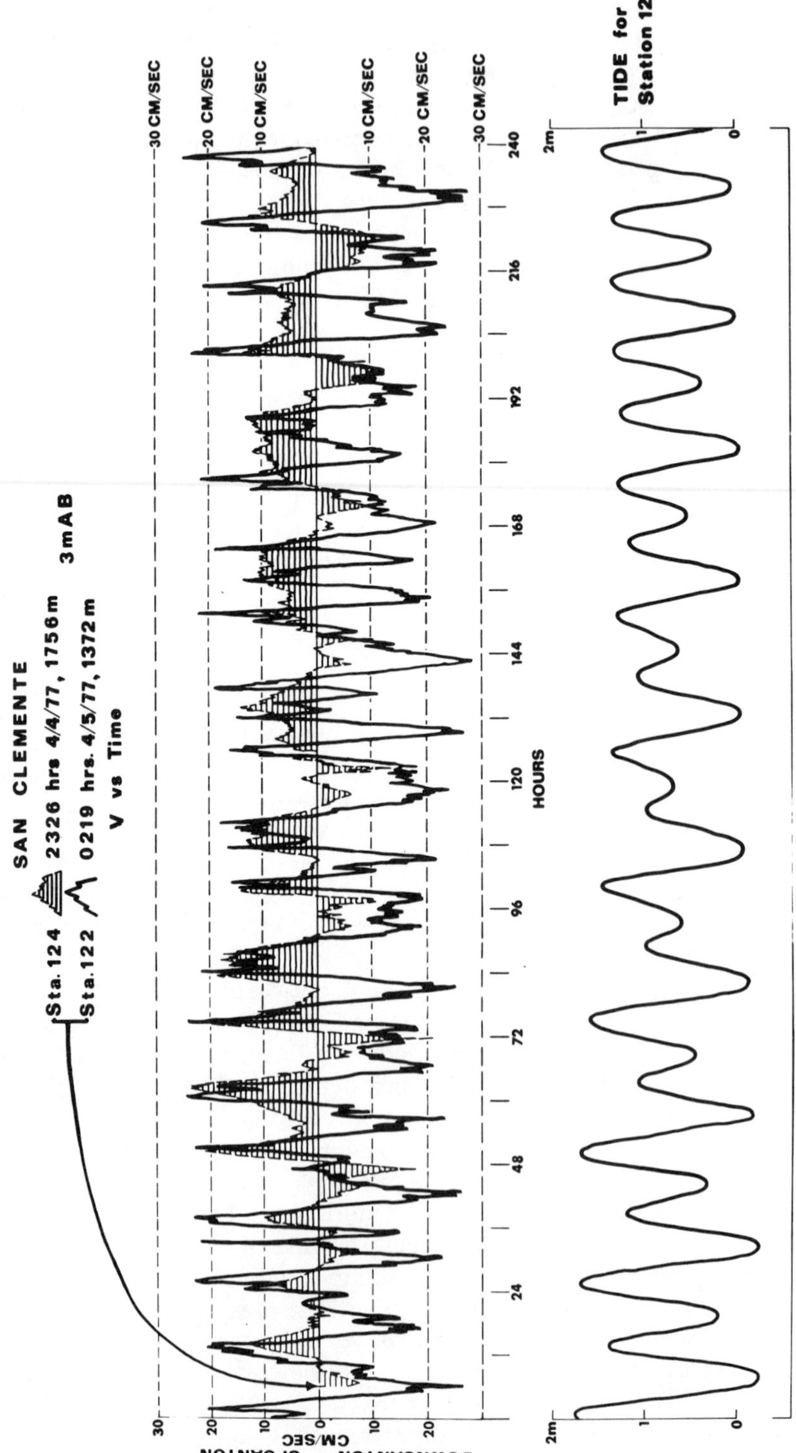

FIG. 69—Superposition of portion of time-velocity curves at 1,372- and 1,756-m depths in San Clemente Rift Valley. The excellent agreement of direction and speed changes between the two curves continued for the remainder of the month-long record showing clearly that the internal waves were moving up the rift valley during that period. These stations were about 12 km apart.

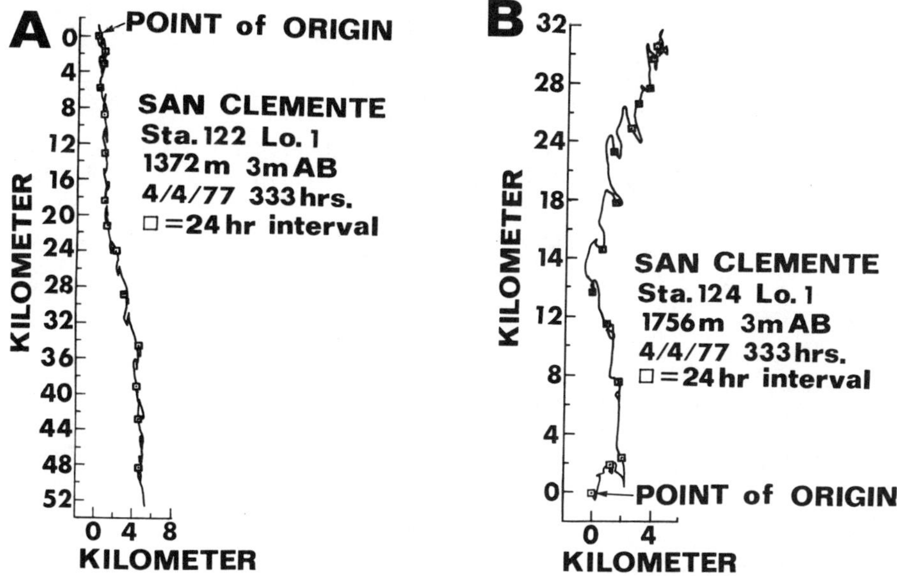

FIG. 70—Progressive vector diagrams of the month-long stations in San Clement Rift Valley. Net flow was downvalley for the shallower station and upvalley for the deeper.

probably broken away from the floats and buoy, either by a violent turbidity current, a sunken log that smashed into the current meter, or by a chain from logging operations. Remnants of the system were recovered later by the Canadian Coast Guard.

The stations in 1976 and 1977 were located at somewhat comparable depths. At the shallow 88-m station in 1976, we had current meters at 3 and 30 m above the bottom. Unfortunately, the 3-m record showed only direction so we have to rely on the 30-m above bottom instrument for speed. In this there is close agreement between the tide phase and current direction so that flow was downvalley when the tide was ebbing and upvalley when the tide was in flood (Fig. 11A). However, the downvalley current was much stronger than that of the upvalley so that net flow was predominantly seaward. There is a good chance that this was not entirely true at the 3-m height because the current flows (with no speed record) lasted about as long upvalley as down. In 1977 our record at 60-m axial depth with the instrument at 3 m above bottom shows a decidedly greater upvalley flow than down (Fig. 11B). This might suggest that a salt wedge was moving landward under the outflowing river water as is found off the Mississippi and some other large rivers (Scruton, 1956; Wright and Coleman, 1971). It is notable that the ebb and flood tides do not correspond as well to the direction of the bottom current in the 1977 record as they did in the 30-m-above-bottom record in 1976. In general the upvalley flows in 1977 occurred during high tide, but at least one occurred during low tide.

The presence of a salt wedge under the outflowing river water is easier to understand in the spring of 1977 because there was less outflowing water due to smaller winter snows, whereas the large flow in 1976 probably pushed out the salt wedge more completely, just as it is driven out of the mouth of the Mississippi during flood stages.

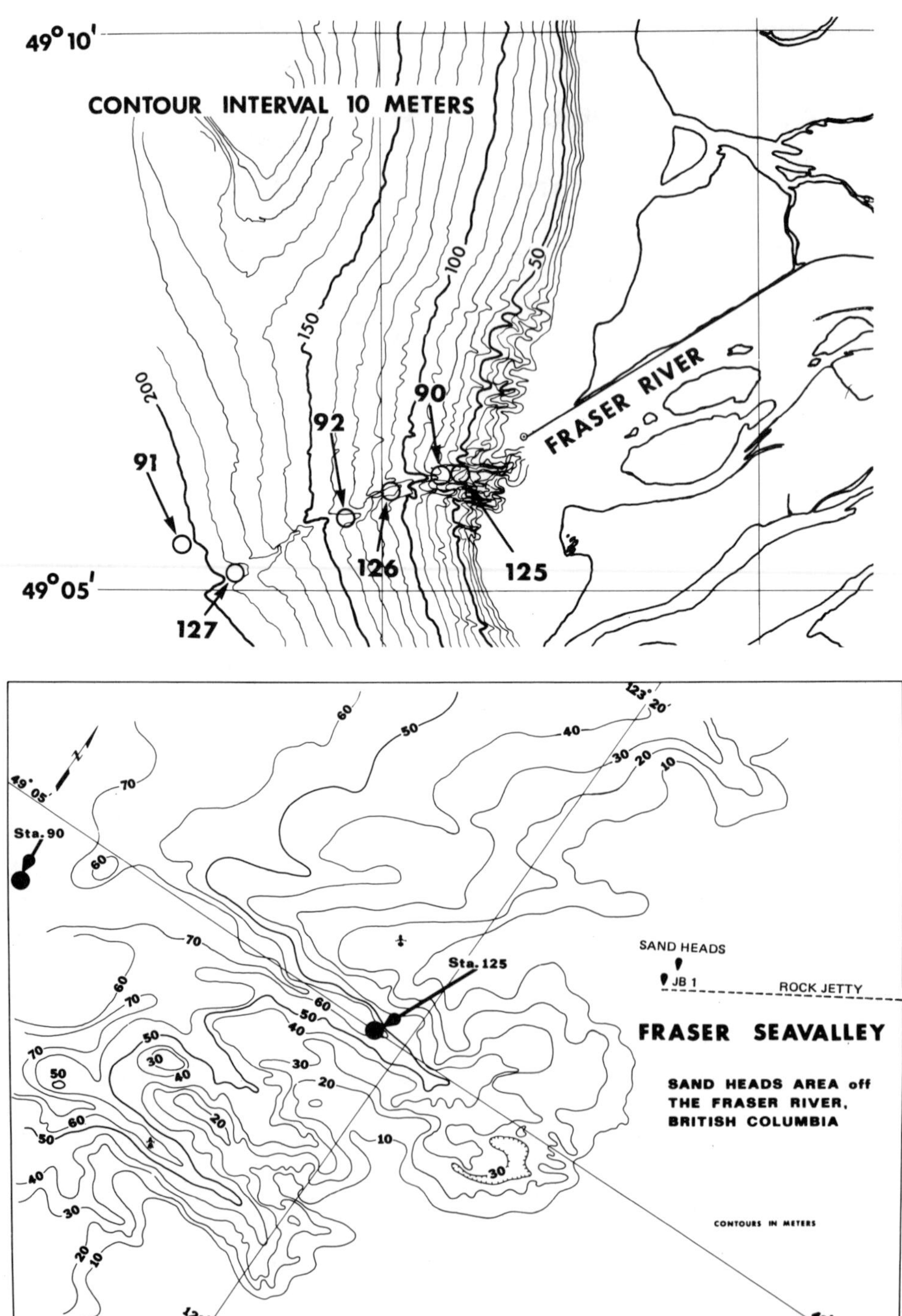

FIG. 71—**A**, contour chart of the slope beyond the Fraser delta in the Strait of Georgia, British Columbia, indicating the Fraser Seavalley. Shows also approximate location of current-meter stations. Contours supplied by J. Luternauer and Department of Energy, Mines, and Resources Canada. **B**, enlargement of the area off the mouth of the Fraser River. Contoured from a survey made by the Public Works Department of Canada. Sounding by Public Works Canada, December 20-22, 1976. Contours in meters.

The middle station (148 m) in 1976 and the 120-m station in 1977 had good records, both at 3 m above bottom; but in the 1976 record only current speed was shown. The 1977 record shows clear tidal periods of alternating direction and the bottom flow was upvalley during falling tides and downvalley during rising tides. Unlike the shallow station of the same period, the net flow was slightly downvalley. Actually the polar plot indicates a northerly direction for downvalley flow and southerly for upvalley flow, possibly due to a bend in the channel or to the general northerly set of the tide. This northerly flow was observed also at the surface during ebbing tide.

At the deep station we had current meters 3 m above the bottom, located at 220 m in 1976 and at 195 m in 1977. Both of these provided good data (Fig. 72). Again both records indicate that the alternating current cycles are related to the tides. In 1976 the currents flowed mostly opposite to the surface tide. However, the strength of the current shows a distinct relation to the tide curve. Thus, all the flows faster than 20 cm/sec occurred during periods when the tide was changing by as much as 3 m and the fastest flow occurred during a 4-m spring tide. The same accordance between fast flows and large changes of tide height was indicated in 1977, and again the tide phase was largely contrary in direction to the current. In both cases the net flow was downvalley although the direction seems to have been more to the northwest than west which is diagonally down the general delta slope. In 1977 the currents were somewhat stronger than during the previous year.

A comparison between flow patterns at the various stations in the two years shows that there is a good match between the shallow and deep stations in 1976 and between the middle and deep stations in 1977 (Fig. 73). The match between the shallow and intermediate stations in 1977 is doubtful, but in all cases the indication was that the internal waves were moving down the valley. In both 1976 and 1977 the internal waves arrived at the deep station about 5 hours after those at the shallow station. This downvalley advance of the internal waves may be related to the large amount of fresh water entering the ocean at the mouth of the Fraser.

The alternation cycle of up- and downvalley flows shows a rather constant increase with depth both in 1976 and 1977. In 1976 at 88 m the average was 10 hours per cycle, and at 220 m it was 12.5 hours. In 1977 at 60 m it was 11.6 hours, at 120 m it was 12 hours, and at 195 m it was 12.7 hours. All of the cycle lengths, however, are clearly related to the tide period.

Mexican Canyons

Cape San Lucas and San Jose Canyons

Around the southern tip of Baja California there are a series of submarine canyons that indent the seafloor and incise the intrusive igneous rocks that make up most of the southern end of that peninsula (Shepard, 1964). At Cape San Lucas the canyon extends into a semiprotected bay (Fig. 74), heading so close to shore that the local fish cannery's short pier extends into the canyon, allowing relatively deep-draft vessels to come in and unload their catch onto the pier. This canyon is located off a large land valley which comes down from the high southern Cape Mountains. Farther east other valleys of the same type are located at the head of submarine canyons, including San Jose, which along with San Lucas was used for our current-meter studies.

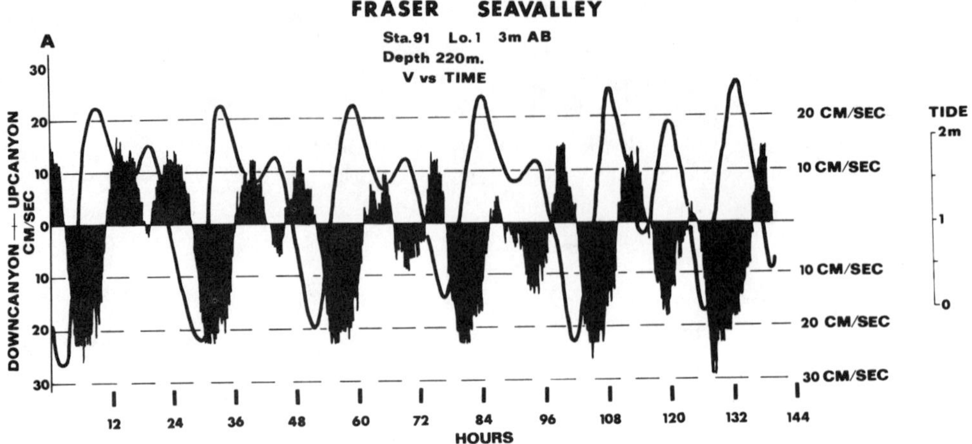

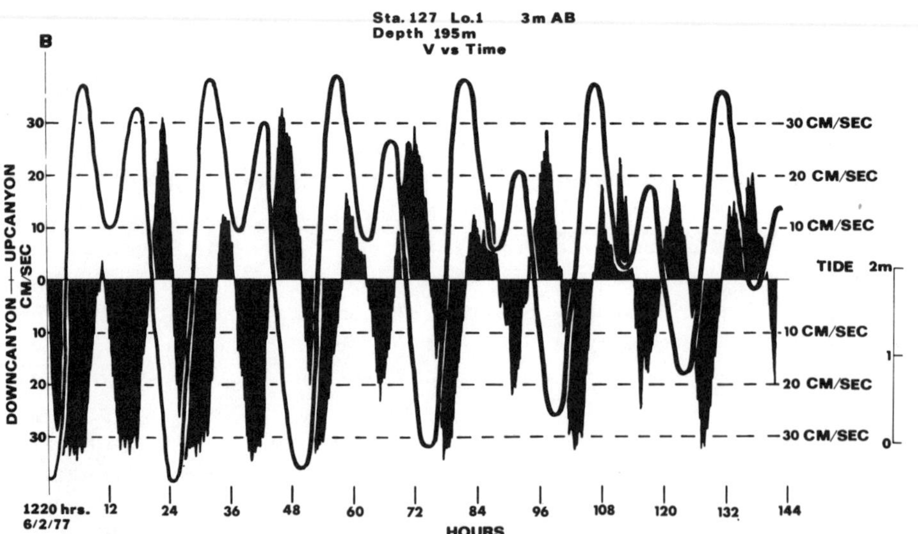

FIG. 72—Comparison between the time-velocity curves at the deep station in the Fraser Seavalley in 1976 (**A**) and in 1977 (**B**). The major flow in both cases is downvalley, but actually more to the north which may be downvalley at these points. Downvalley flows, in both records, are in general related to times of rising tide.

San Lucas Canyon is protected from the large waves of the open ocean although occasionally strong winds from the east blow into this bay, setting up steep short-period waves. However, our current-meter studies were not operated during such storms.

Dives into San Lucas Canyon in Cousteau's *Diving Saucer* and in Westinghouse's *Deepstar 4000* have given us a good idea of the character of the canyon floor and walls (Shepard and Dill, 1966, p. 105-114). The head of the canyon has a consistently muddy bottom, but the canyon becomes sandy beyond depths of about 300 m. Abundant ripple marks partly covered by recently fallen granite fragments have been seen at this greater depth.

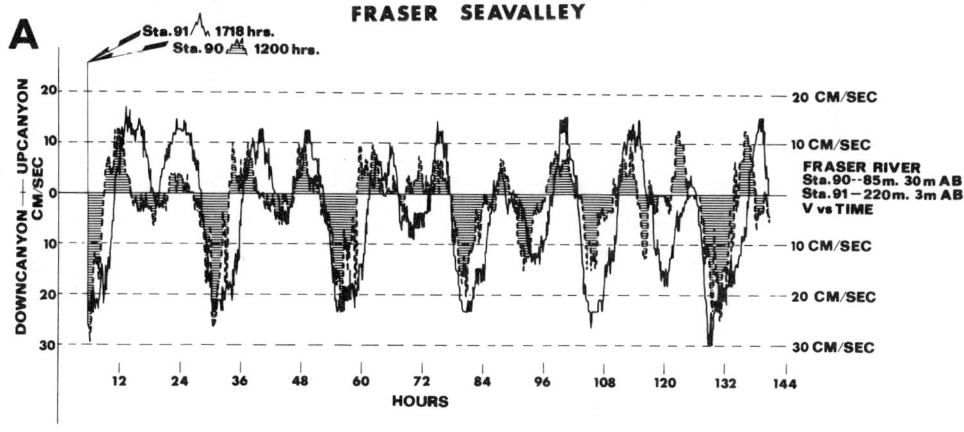

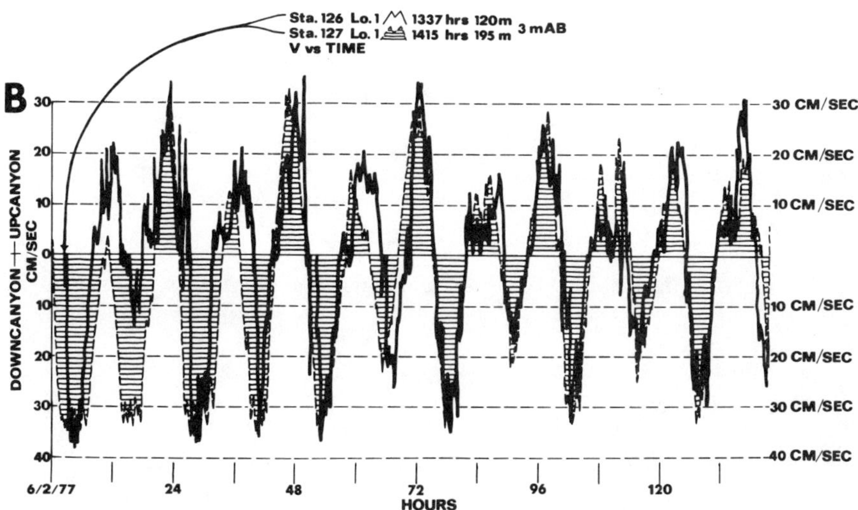

FIG. 73—Superposition of time-velocity curves at the shallow and deep stations in 1976 (**A**) and in 1977 (**B**). Arrivals in both cases are later downvalley indicating internal waves are advancing in that direction, probably because of the large flow of water coming into the system from the Fraser River.

In 1971 we had three stations in San Lucas Canyon and obtained five current-meter records, all at 3 m above bottom. These were located in the canyon axis at 137, 216, and 328 m. The currents at the 137-m and 216-m stations were slower than at 328 m. One of the two 137-m stations showed currents up to 18 cm/sec (Fig. 75), but the maximum current at the 216-m station was 7 cm/sec in the downcanyon direction (Fig. 75). On the other hand, the deep station had downcanyon flows as fast as 37 cm/sec and an upcanyon maximum of 18 cm/sec (Fig. 75B).

Unfortunately in this early operation our current meters were set to indicate each speed mark only after every 64 rotations of the rotor. With slow currents the time-velocity curves at the shallow stations do not show as many of the alternations of

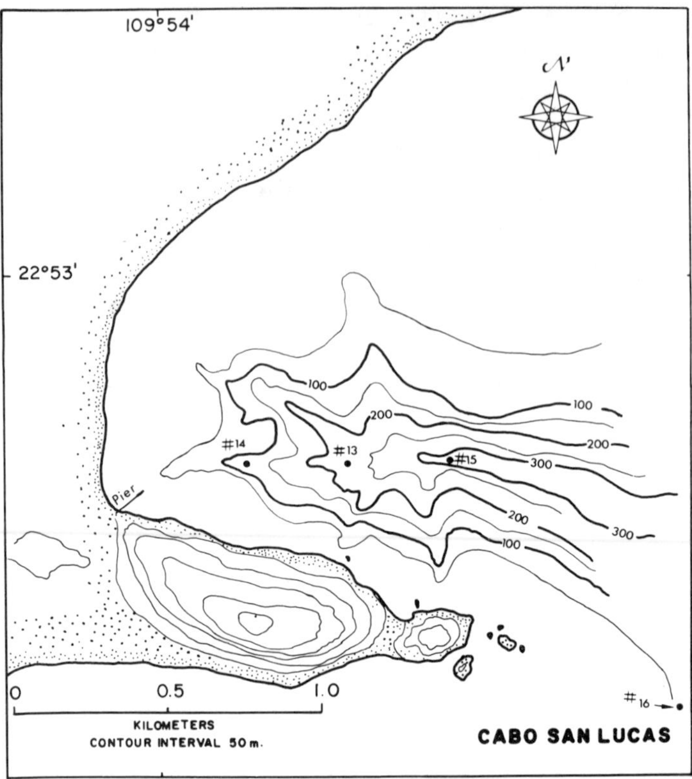

FIG. 74—Contour chart of the head of San Lucas Canyon at the tip of Baja California with the location of the current-meter stations included. Based on surveys by Shepard (1964).

up- and downcanyon currents as are indicated in the continuous direction record of the tape. Therefore we can tell much better how often the currents reversed by using a copy of the tape (Fig. 76), rather than by using the time-velocity curves. Using the former, the average time of a cycle for the 137-m station is 1 hour and 5 minutes; for the 216-m station it is about 2 hours, and for the 328-m station it is about 3 hours. Thus the alternation period increases steadily with depth. The time-velocity plots failed to indicate this sequence due to the part with slow currents, as explained above.

We do not have any stations deeper than 328 m in San Lucas Canyon, but we have two records of relatively short duration in the adjacent San Jose Canyon at depths of 1,408 and 1,860 m.[4] Although these measurements were taken for only a short period, it is clear that the up- and downcanyon alternations are, timewise, much farther apart. At 1,408 m the cycles are of the order of 25 hours, representing the diurnal tide (Fig. 75). On the other hand at the 1,860-m station the alternations are related to the semidiurnal tide with a 12.5-hour cycle. No reason for this change

[4]The reason for choosing San Jose Canyon instead of San Lucas for the deep records was that permission had not yet been obtained from Mexico to work inside the 12-mile limit when we returned to the San Lucas area in 1975, so we had to choose two stations in the adjacent canyon that were beyond the limit.

from diurnal to semidiurnal tide period is evident. However, it seems likely that the period of alternation becomes essentially tidal at a depth around 1,000 m in the Cape San Lucas area where the tides probably are of the order of 1 m or less at the canyon head.

Rio Balsas Delta Area, West Mexico

Rio Balsas is the second largest west coast river in Mexico. Some 250 km north of Acapulco, it has built a delta into the Pacific, which has covered the continental shelf and now is building out onto the continental slope. Here, as in all such cases, valleys (here of canyon proportions) have developed on the advancing slope (Fig. 77). These valleys apparently are due to a combination of levee building on the valley sides and excavation into the advancing delta mass. This was clearly shown by seismic profiles run along the slope (Reimnitz et al, 1975). The principal valley, called Rio Balsas Canyon, is located off the main river mouth and can be traced down the slope to a depth of at least 2,000 m where soundings are too scarce to trace it deeper. To the east one large tributary, Petacalco Canyon, heads off an old abandoned land channel. This extends southeast for 25 km and then bends to the west-southwest, joining the main valley at a depth of about 1,300 m. Other valleys cut the slope to the west of Rio Balsas Canyon but apparently do not join the main valley.

The study of the currents off the Rio Balsas delta was initiated by Reimnitz (1971). While scuba diving in a small tributary of Rio Balsas Canyon during a period of large swell he encountered intermittently flowing turbidity currents. At the same time as these downcanyon pulses, surface boats observed rip currents flowing seaward (see also Reinmitz et al, 1975).

In 1975 we returned to the area in a joint expedition with Reimnitz and others of the U.S. Geological Survey and obtained current-meter measurements at five stations along Rio Balsas Canyon and at one station near the head of neighboring Petacalco Canyon. In Rio Balsas Canyon we placed our instruments 3 m above the bottom at axial depths of 285; 384; 658; 1,290; and 1,904 m with an extra instrument at 30 m above bottom at 658 m. In Petacalco Canyon at 110 m we also had current meters at 3 and 30 m above bottom.

One of the interesting things about the currents off the Rio Balsas delta is that the alternation period of up- and downcanyon flows at 3 m above bottom shows a continuous lengthening of the reversal cycles with depth. Thus at 110 m in Petacalco Canyon the average flow cycle was 3.6 hours (4 hours at 30 m above the bottom); at 285 m in Rio Balsas Canyon it was 5.7 hours; at 384 m it was 6.0 hours; at 658 m it was 7.2 hours (8 hours at 30 m); at 1,290 m it was 10.2 hours; and at 1,904 m it was 12.7 hours (Fig. 77). The period of current flow becomes tidal (12.4 hours) at some depth between 1,290 and 1,904 m, which is in accord with other areas that have less than 1-m tide range. The tide record showed diurnal tides at the beginning, becoming semidiurnal and finally diurnal again at the end. However, the currents are entirely related to semidiurnal tides at the 1,904-m depth (Fig. 7E).

The two cases where we have records at both 3 and 30 m above the bottom have a strange contrast. At the 658-m station in Rio Balsas Canyon the currents at the two heights are very similar, changing directions at about the same time and having similar velocities (Fig. 19A). On the other hand at the 110-m station in Petacalco Canyon the two heights show very contrasting currents (Fig. 19B). In fact, along

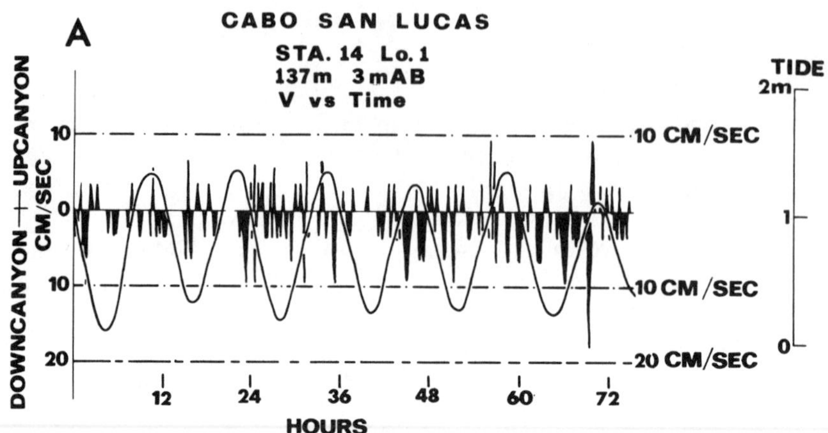

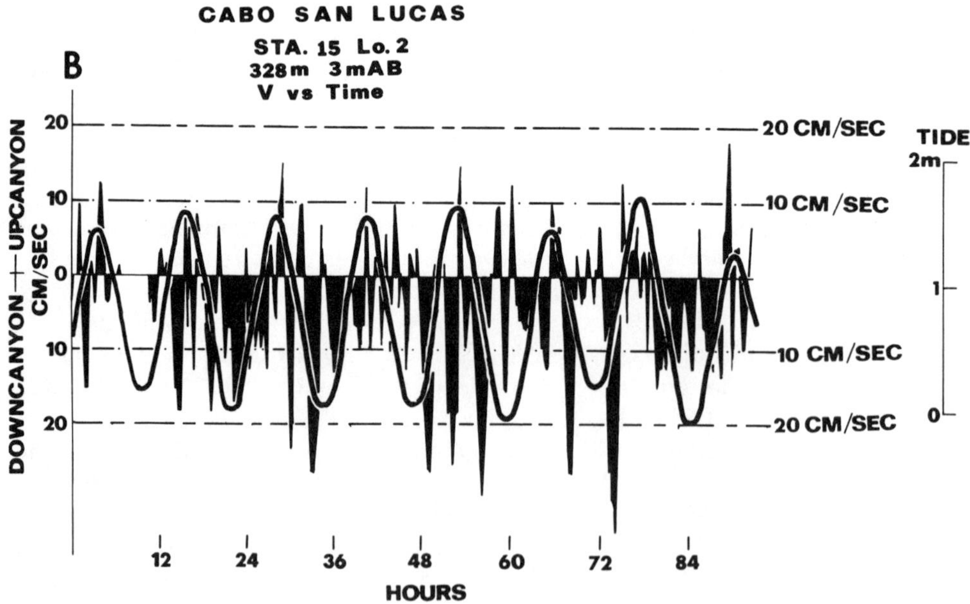

FIG. 75—Time-velocity curves from stations in San Lucas Canyon (**A, B**) and neighboring San Jose Canyon (**C, D**) show high frequency of alternations in shallow water with the small tides and the currents with tidal frequency for the deep-water stations.

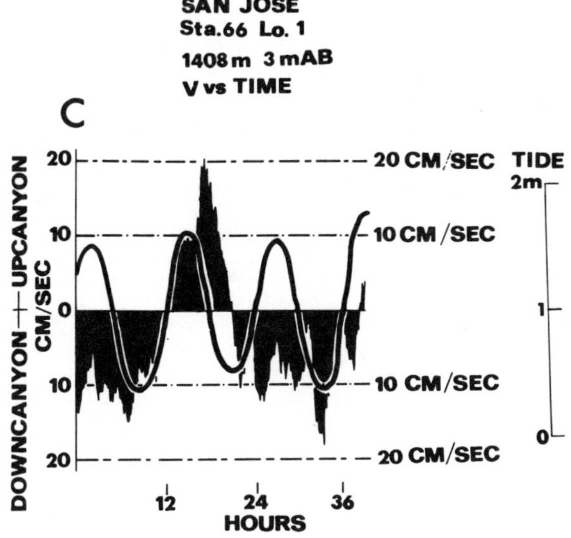

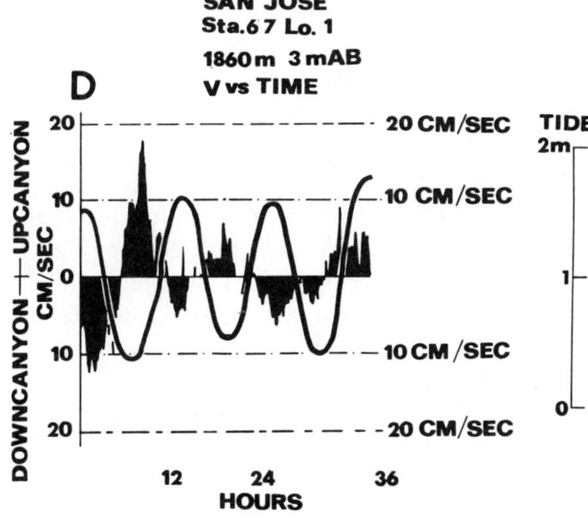

FIG. 75–(continued)

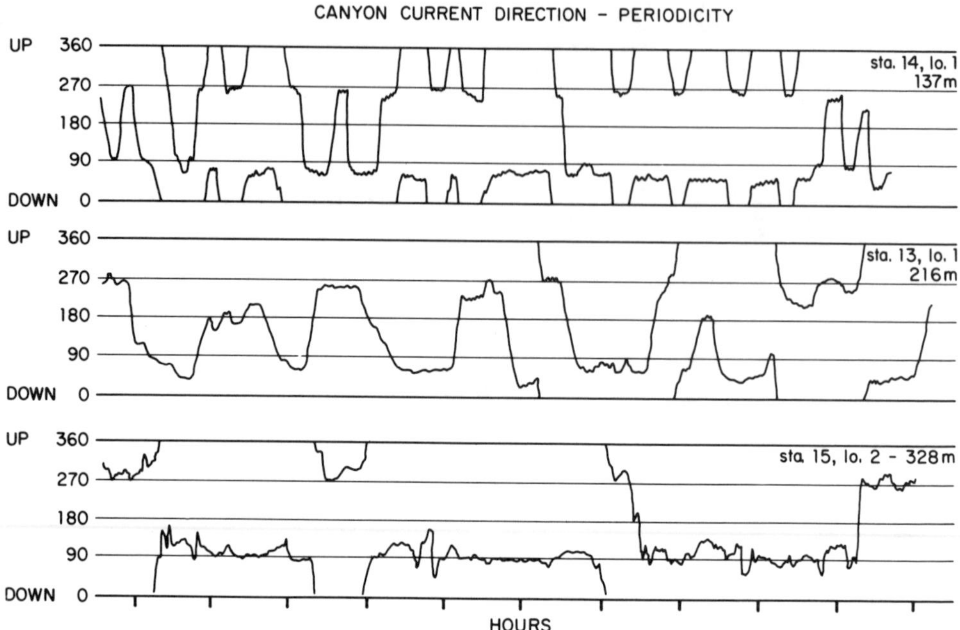

FIG. 76—Copy of portions of current-meter records from the three shallow stations at the head of San Lucas Canyon showing the high frequency of direction alternations at the two shallower stations and longer periods at the deeper station. Such rapid alternations are not shown by the usual time-velocity curves where 64 revolutions are used as they were here.

much of the Petacalco records, the currents are shown as flowing in opposite directions. However, some of the strong flows at one height are nearly matched in time by a strong flow in the same direction at the other height.

At Rio Balsas the net flow is always downcanyon except at the Petacalco Canyon station where it is upcanyon for both the 3 and 30-m above bottom levels. Here again we find upcanyon net flow at a shallow canyon station as in many West Coast canyons.

Comparing time-velocity curves at adjacent stations, a phase calculation between the 384-m station and the 658-m station for waves in the 6 to 10-hour period range shows that the wave form is moving downcanyon at an average speed of approximately 51 cm/sec.

A comparison in this area between stations at 1,290 m and 1,904 m (Fig. 78) is not at all conclusive, as the 1,290-m station was of short duration. Here the indications are for downcanyon advance of internal waves, but the time difference is very short for the 18-km distance, indicating a speed much faster than any other comparison we have found.

Turbidity currents at Rio Balsas—Because Reimnitz previously observed turbidity currents off the Rio Blasas delta, it was not a great surprise that our current-meter records included another example (Shepard et al, 1976). We set a current meter at 285 m in the axis of Rio Balsas Canyon about 12 hours before we noted that swells coming into the coast from a distant storm were rising well above what we observed previously. This period of high swells lasted for about 40 hours. When we recovered

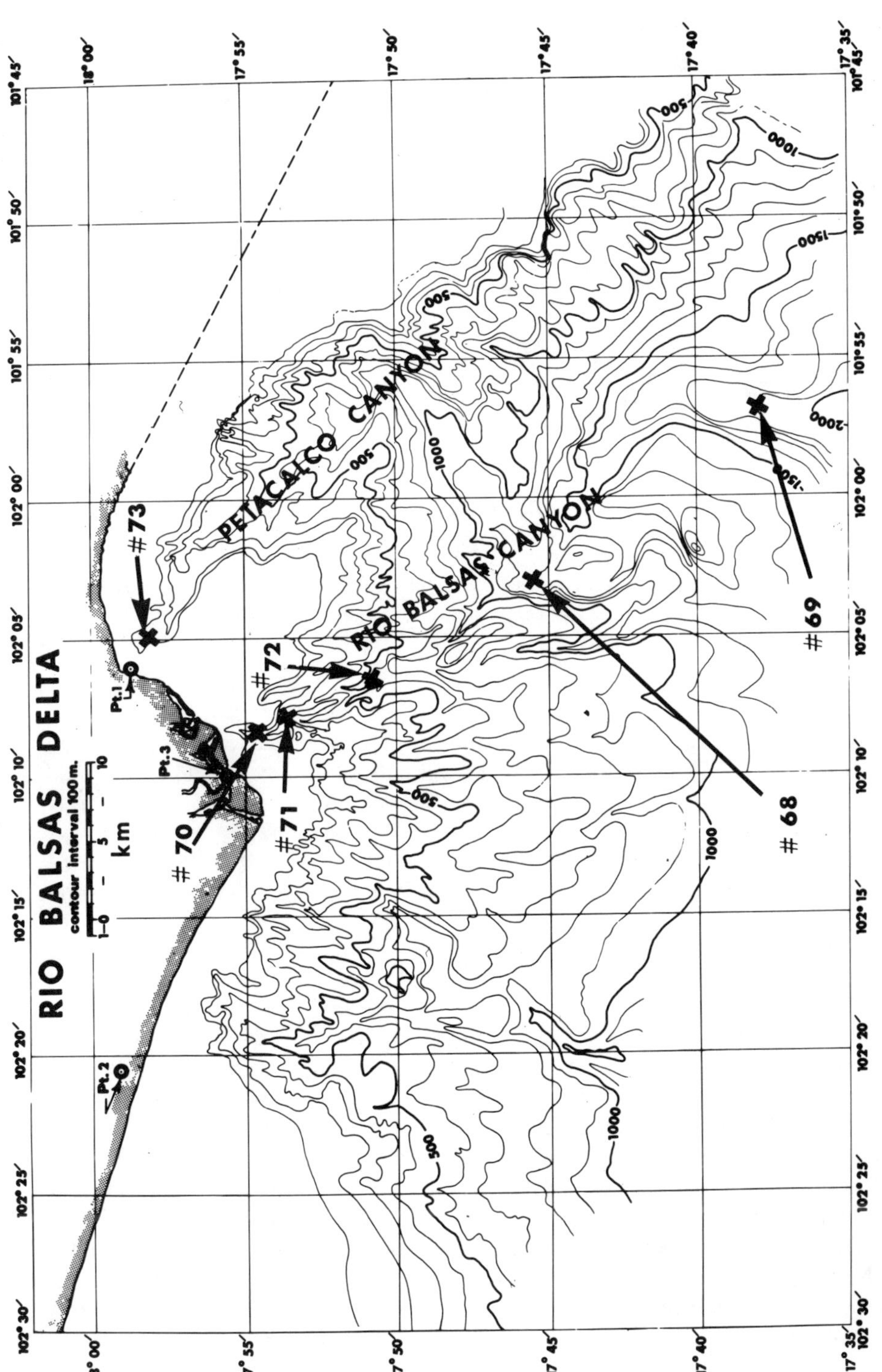

FIG. 77—Rio Balsas delta and the canyons and seavalleys that incise the submarine slope beyond. Locations of current-meter stations are shown. Petacalco Canyon apparently changes course at a depth of 900 m and joins Rio Balsas Canyon at about 1,200-m depth. Based on surveys by Erk Reimnitz and F. P. Shepard with U.S. Geological Survey equipment in 1975.

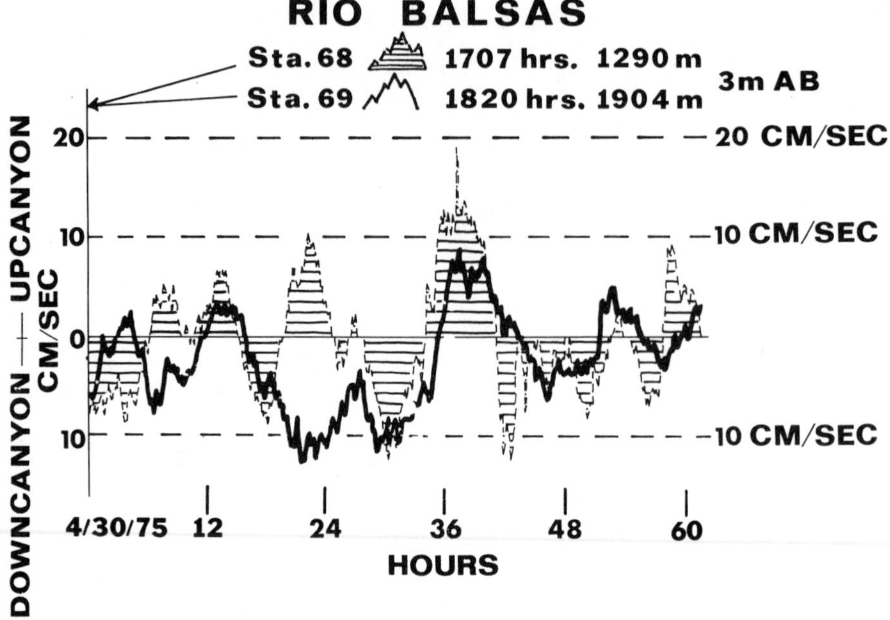

FIG. 78—Superposition of time-velocity curves at 1,290- and 1,904-m depths in Rio Balsas Canyon. Times of best fit are later at the deeper station suggesting downcanyon advance of internal waves.

the current meter, we found that during this period there was a series of strong downcanyon pulses which occurred close to the time of the relatively large diurnal high tides (Figs. 22, 23). The pulses reached a velocity of 73 cm/sec. There seemed to be little doubt that they were a form of turbidity current and almost certainly were related to the high swell as had been reported previously by Reimnitz. The fact that the second group of 6 pulses occurred somewhat ahead of the high tide may mean simply, that the high tide occurred earlier in the canyon head than up the estuary of the river where our tide gauge was located.

East Coast Canyons

Off the East Coast of the United States there is a series of 26 canyons[5] that cut the continental slopes (Fig. 79), most of them penetrating into the edge of the continental shelf but not extending as far as the coast. Ten of them, located off Georges Bank and Nantucket Shoals, have no apparent connection with mainland rivers. The others also are relatively independent of the mainland except for Hudson Canyon, which is essentially a seaward continuation of a shallow shelf valley on the inside that extends into the New York Bight and hence into the estuary of the Hudson. Other canyons farther south apparently have small shelf valleys pointing towards Delaware and Chesapeake Bays. These undoubtedly connected with the canyon heads when the sea level was lowered during glacial stages. However, the distributaries from the Hudson may have also extended this far south according to Kelling et al (1975).

[5]One hundred ninety canyons including eastern Canada, according to Emery and Uchupi (1972).

The current-meter measurements were made possible in this area on the NOAA ship *Researcher* under the direction of George Keller (Keller and Shepard, 1978).

Hydrographer Canyon

Among the canyons off Georges Bank, Hydrographer is the largest and penetrates farthest (14 km) into the continental shelf (Fig. 80). This canyon has been investigated during many descents in deep-diving vehicles and has been extensively dredged and cored by geologists particularly from Woods Hole Oceanographic Institution (Stetson, 1936, 1938, 1949; Emery and Uchupi, 1972). It is a narrow canyon with steep (locally vertical) walls. The samples show that sand is common. Ripple marks and rock bottom have been seen in many bottom photographs and observed from deep-diving vehicles, suggesting relatively strong currents.

The currents in Hydrographer Canyon were measured at axial depths of 348, 512, and 713 m. The instruments were at 3 m above bottom at all stations and additional instruments were at 30 m above bottom at 348 and 512 m. Time-velocity curves from stations show close relations between the tides and alternation cycles (Fig. 16). However, the average cycle, as in so many canyons, shows a slight increase in period with depth. At 348 m it is 11.5 hours; at 512 m it is 12 hours; and at 713 m it is 13.0 hours.

The most surprising feature about the progressive vector diagram is that the net flow for the three stations at 3 m above bottom shows that the shallow and deep stations had a net flow downcanyon and the middle station clearly had a net flow upcanyon (Fig. 81A-C). The middle station also had net flow upcanyon at 30 m above the floor. At the shallow station at 30 m above bottom the net flow is cross-canyon with equal up- and downcanyon flows (Fig. 81D).

Examining the relation of the tide curves to the time-velocity curves of the three stations (Fig. 16), one can see that the peak up- and downcanyon flows shift progressively later timewise in relation to the tide[6] in the upcanyon direction. This also is shown by matching the curves (Fig. 82). The advance of internal waves seems unusually fast in this canyon, as there is a difference of only 1.5 hours between the deep and shallow station, a distance of about 9 km. The match is good, but it would also be almost as good if one matched these tidal-dominated curves with positions at the next tidal interval. The chief arguments that the matching used is correct are that they show the best agreement for the peculiarities of the curves at the three stations, and that flow peaks occur progressively earlier in relation to depth (compare tide).

The velocities of the flows are amazingly high at the shallow station where speeds of 52 cm/sec occurred in a downcanyon direction. The upcanyon flows in the middle station show velocities in excess of 40 cm/sec. These are among the fastest currents of any of our surveys, excepting turbidity currents. Still higher upcanyon and downcanyon velocities have been observed from deep-diving vehicles (B. C. Heezen, personal commun.). The high velocities are in keeping with the sandy and rocky character of the bottom.

Cameras placed on the bottom of Hydrographer Canyon at 710 m during the time of the current-meter survey took time-lapse photographs. The photographs showed

[6]The tide used was from Georges Shoal where the tide is almost certainly of smaller amplitude than on the continental slope, just as it is in a narrow bay entrance like the Golden Gate.

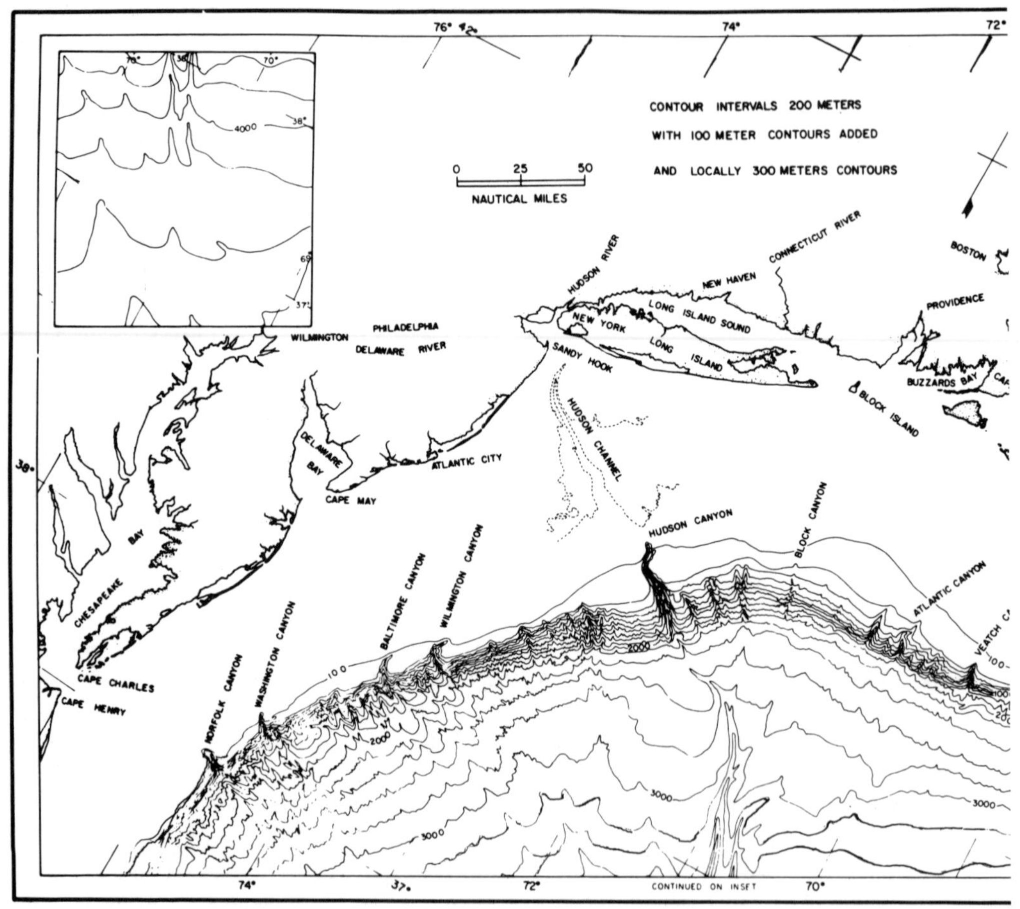

FIG. 79—The canyons off the northeastern United States. From map by E. Uchupi, U.S. Geological Survey.

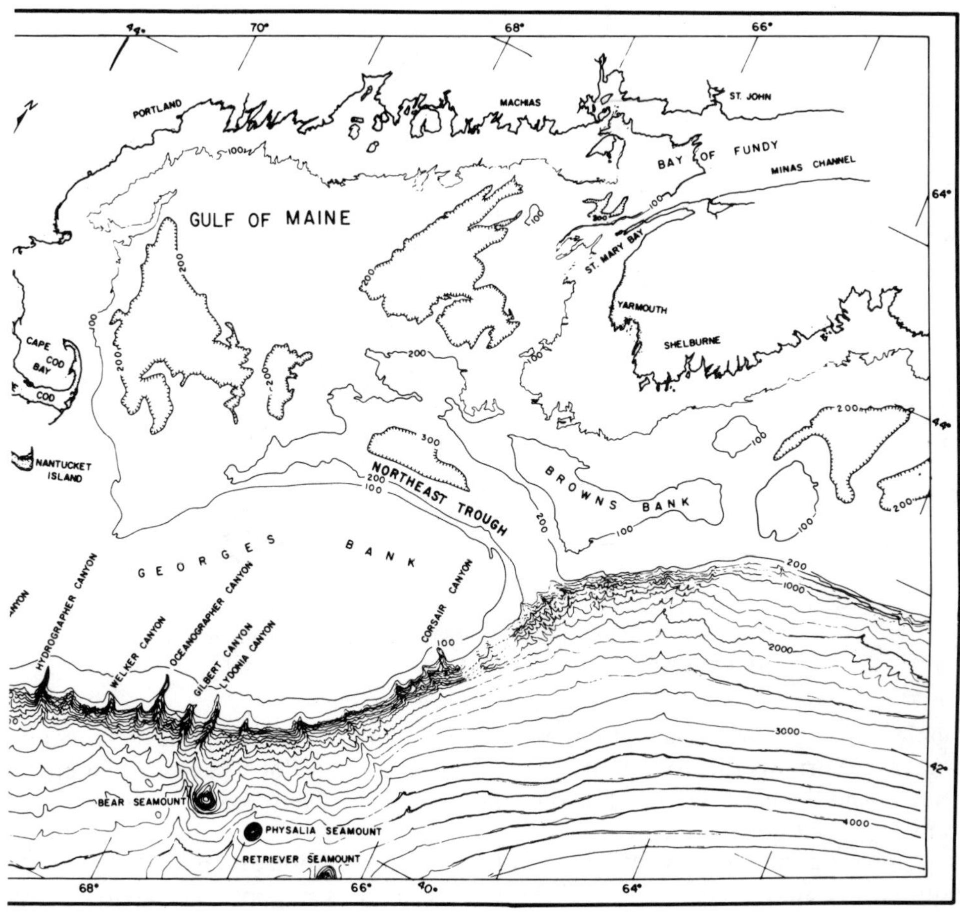

PORTLAND
MACHIAS
ST. JOHN
BAY OF FUNDY
MINAS CHANNEL
GULF OF MAINE
ST. MARY BAY
YARMOUTH
SHELBURNE
CAPE COD BAY
NANTUCKET ISLAND
NORTHEAST TROUGH
BROWNS BANK
GEORGES BANK
HYDROGRAPHER CANYON
WELKER CANYON
OCEANOGRAPHER CANYON
GILBERT CANYON
LYDONIA CANYON
CORSAIR CANYON
BEAR SEAMOUNT
PHYSALIA SEAMOUNT
RETRIEVER SEAMOUNT
70°
68°
66°
64°
42°

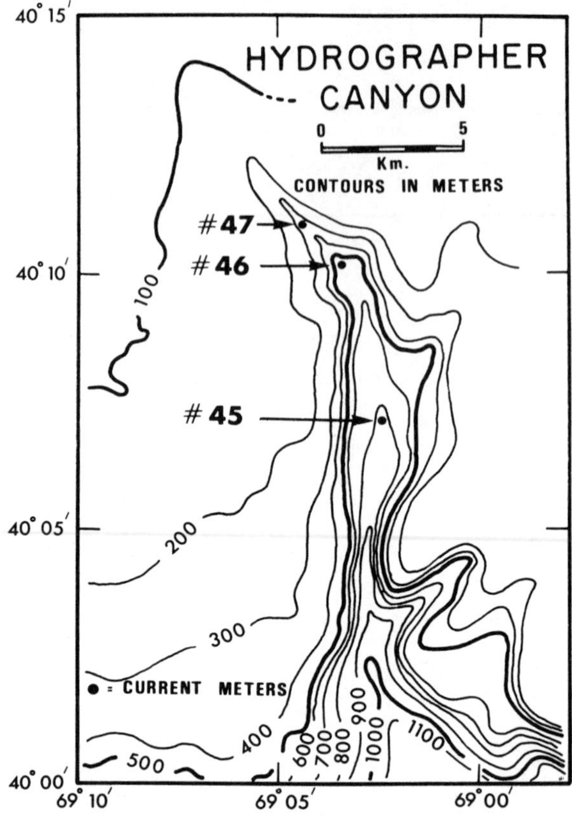

FIG. 80—Contour chart of Hydrographer Canyon with current-meter station locations.

ripple marks in the sand bottom which apparently were migrating downcanyon whenever currents (measured 100 cm above the canyon floor) exceeded 26 cm/sec (Keller and Shepard, 1978). Thus the transport of relatively coarse sediment in this canyon is well documented.

Comparing the time-velocity curves of the shallow station at 3 m and 30 m above bottom, one finds that the time of direction change is in good agreement, but the peak downcanyon speeds at several points on the 3-m curve correspond to slower velocities on the 30-m curve (Fig. 83). On the other hand, the upcanyon peak speeds generally are similar. At the intermediate (512-m) station the time agreement between the two curves is not as good, and the strong upcanyon flows at 3 m correspond to slower speeds at 30 m above the bottom.

The polar plots of the three stations at the 3-m level are somewhat puzzling, as they do not agree with the supposed north-south direction of the canyon axis (Fig. 84). Only the shallow station relates clearly to the up- and downcanyon direction with a NNE and SSW axis, and even this contrasts with the supposed northwest-southeast trend shown on the map for Stations 46 and 47 (Fig. 80). The large amount of crosscanyon direction at the intermediate and deep stations may be due to the location of the meters at curves in the canyon axis such as we found at Station 32 in Carmel Canyon.

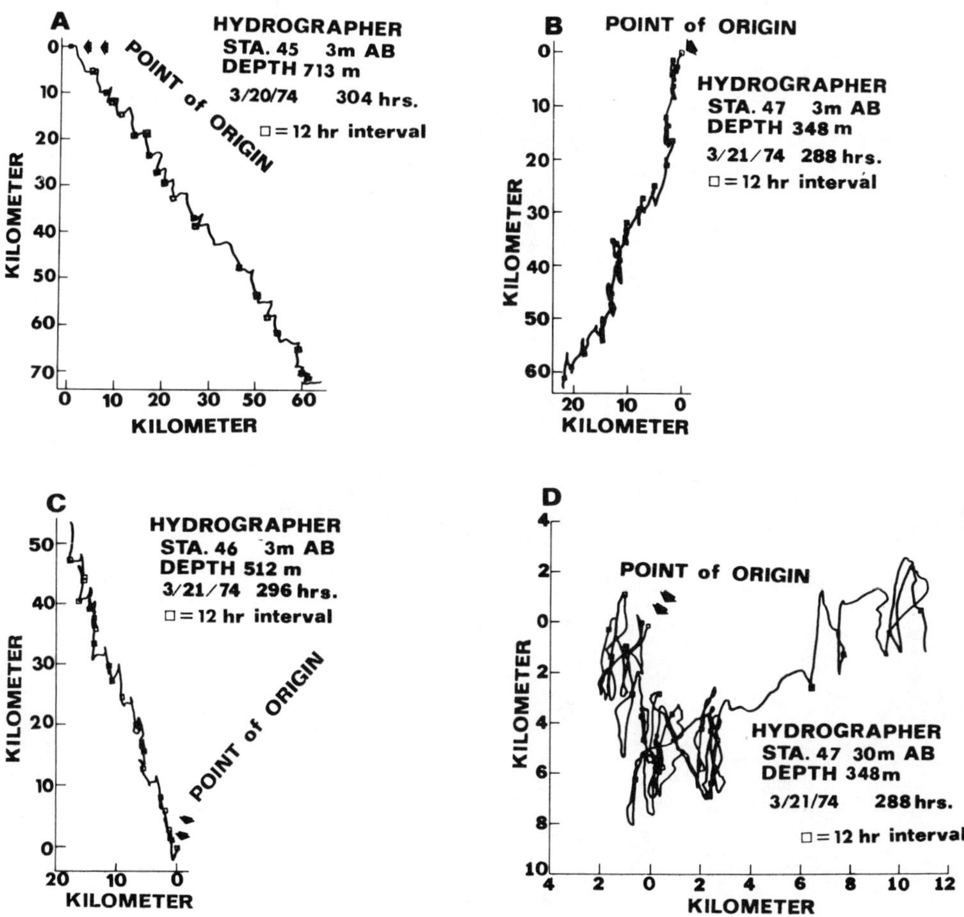

FIG. 81—Progressive vector diagrams of the records from Hydrographer Canyon. Note the contrast of net flow direction at mid-depth station (**C**) from the others. Shows also a complex flow direction at 30 m above bottom at the shallow station (**D**).

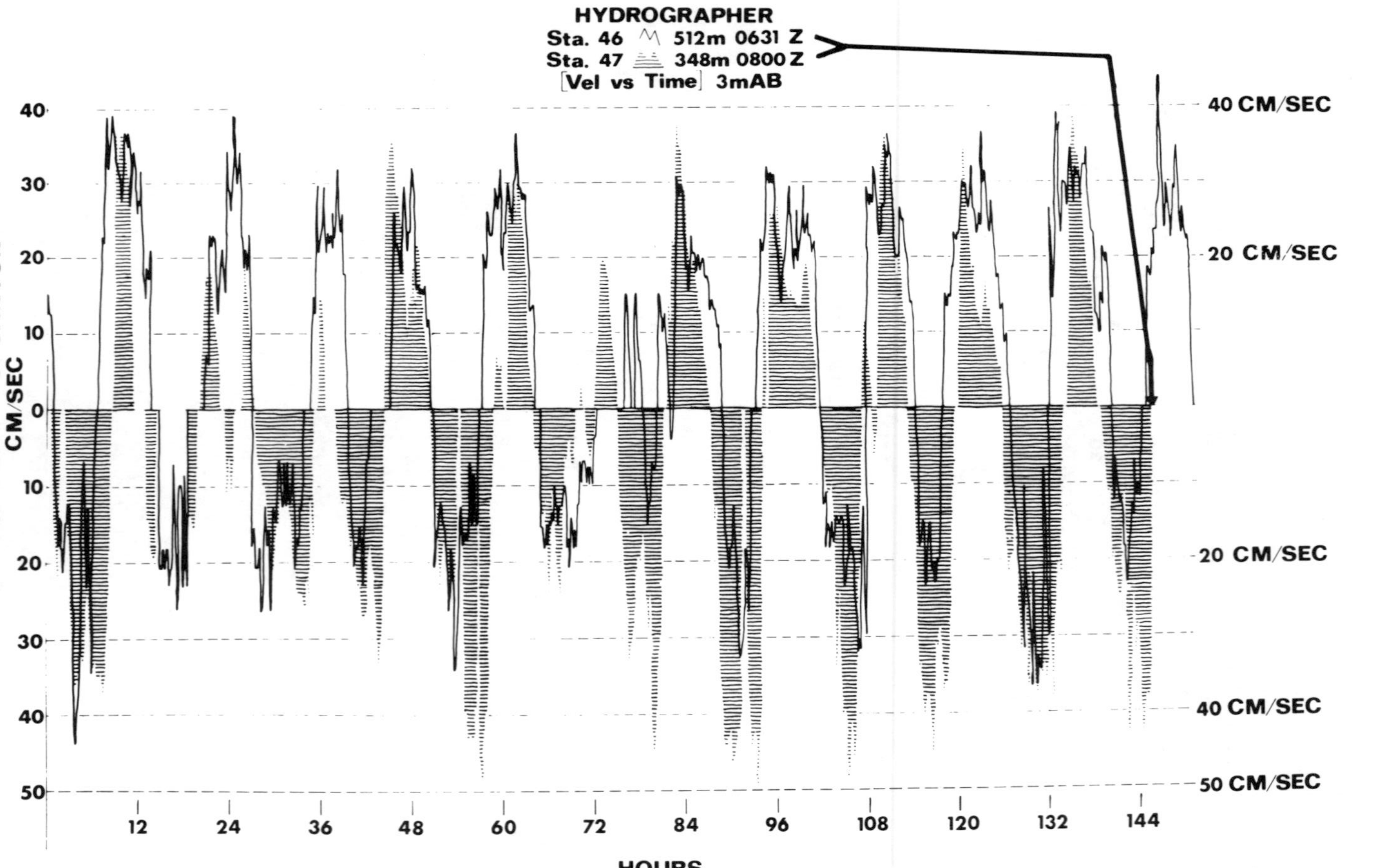

FIG. 82—Superposition of time-velocity curves for records at 348- and 512-m depths inn Hydrographer Canyon. The later times for the shallow station, shown by the best fit, indicate upcanyon advance of the internal waves. Same conclusion can be drawn from Figure 16 where peak flows show a progressive relation to tides at the three Hydrographer Canyon stations.

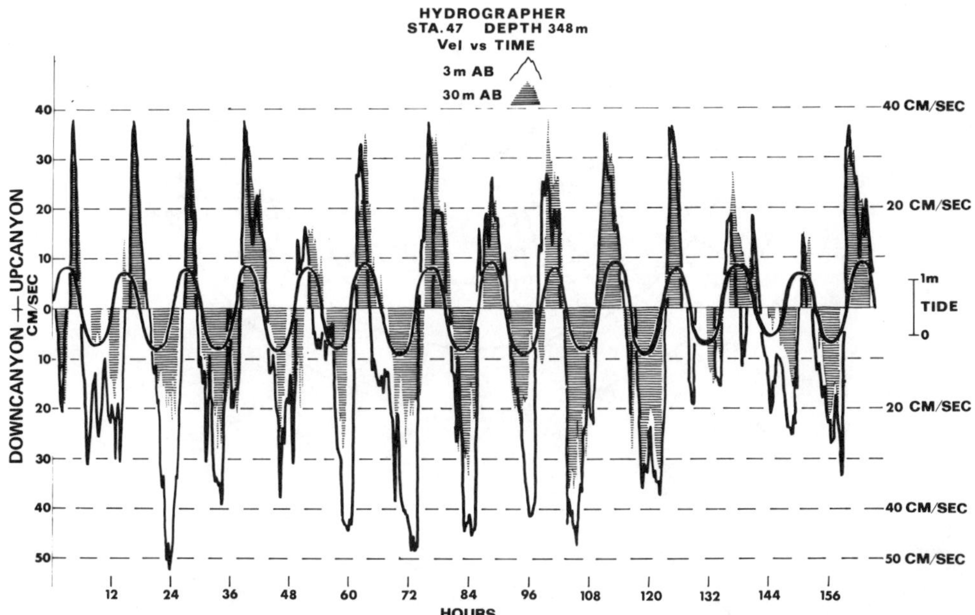

FIG. 83—Overlay of time-velocity curves at 3 and 30 m above bottom at 348 m in Hydrographer Canyon. Upcanyon flows show good agreement in time and speed but the downcanyon flows although they agree rather well in time have far slower speeds at 30 m than at 3 m.

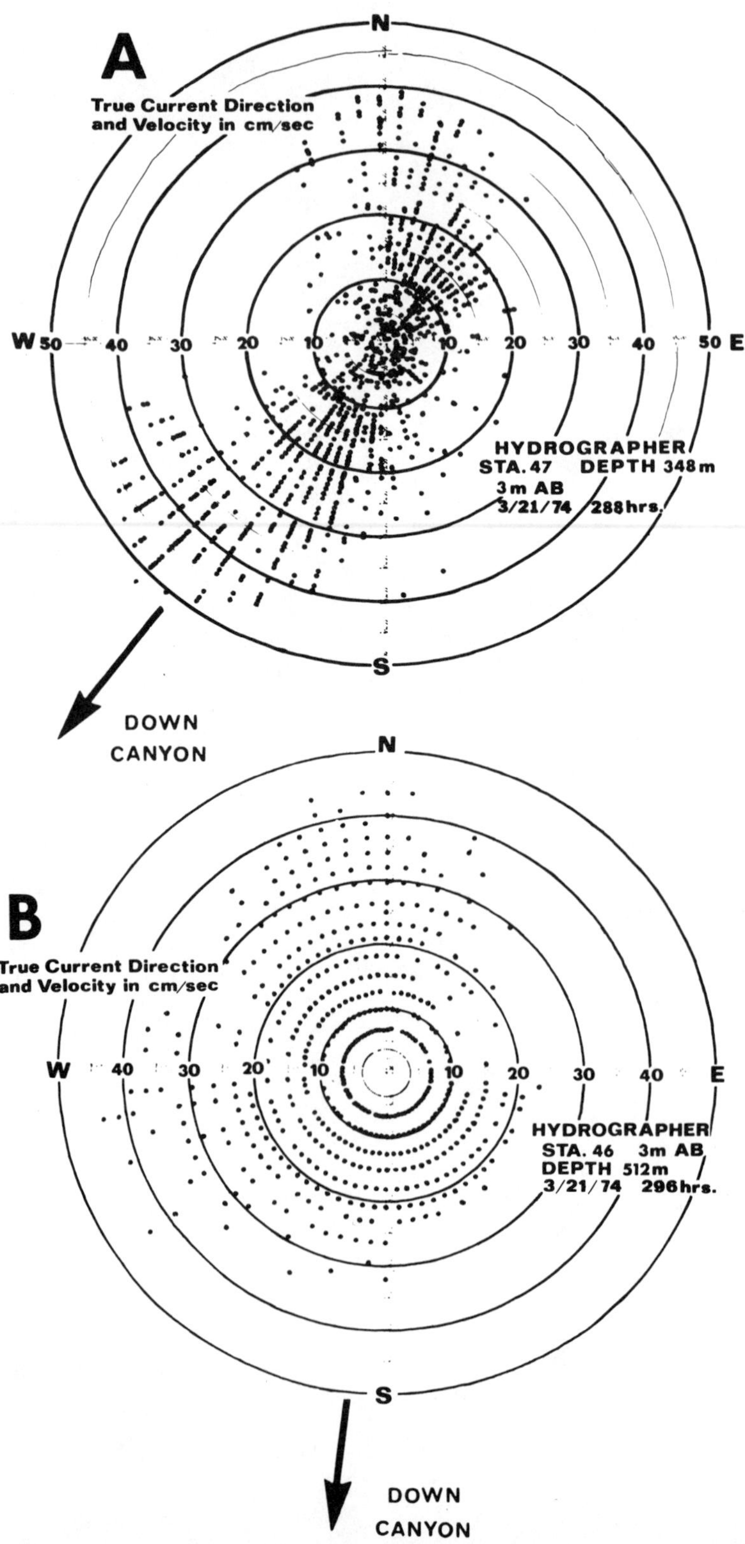
A
True Current Direction
and Velocity in cm/sec
N
W 50 40 30 20 10 10 20 30 40 50 E
S
HYDROGRAPHER
STA. 47 DEPTH 348m
3m AB
3/21/74 288hrs.
DOWN
CANYON
B
True Current Direction
and Velocity in cm/sec
N
W 40 30 20 10 10 20 30 40 E
S
HYDROGRAPHER
STA. 46 3m AB
DEPTH 512m
3/21/74 296hrs.
DOWN
CANYON

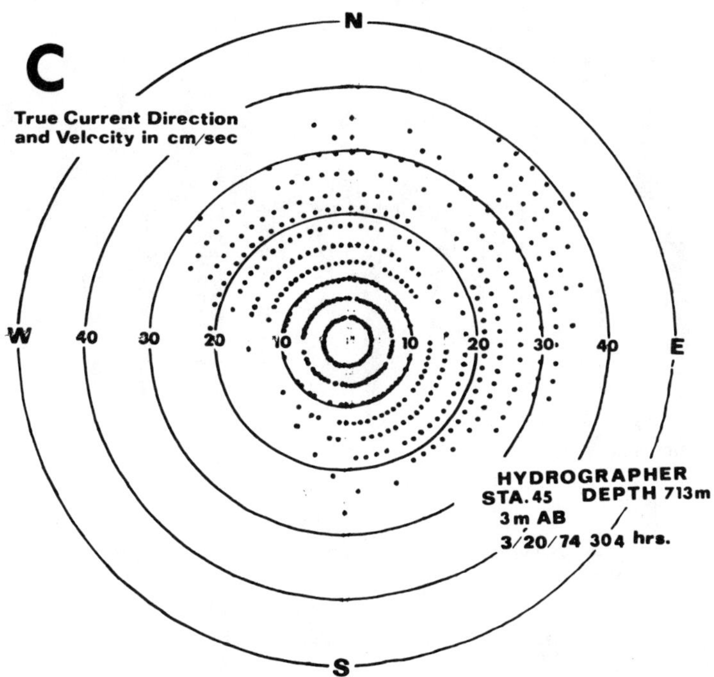

FIG. 84—Polar plots of the three stations in Hydrographer Canyon at 3 m above bottom. Only at the shallow station were the currents consistently flowing close to what appears to be an axial direction (Fig. 80).

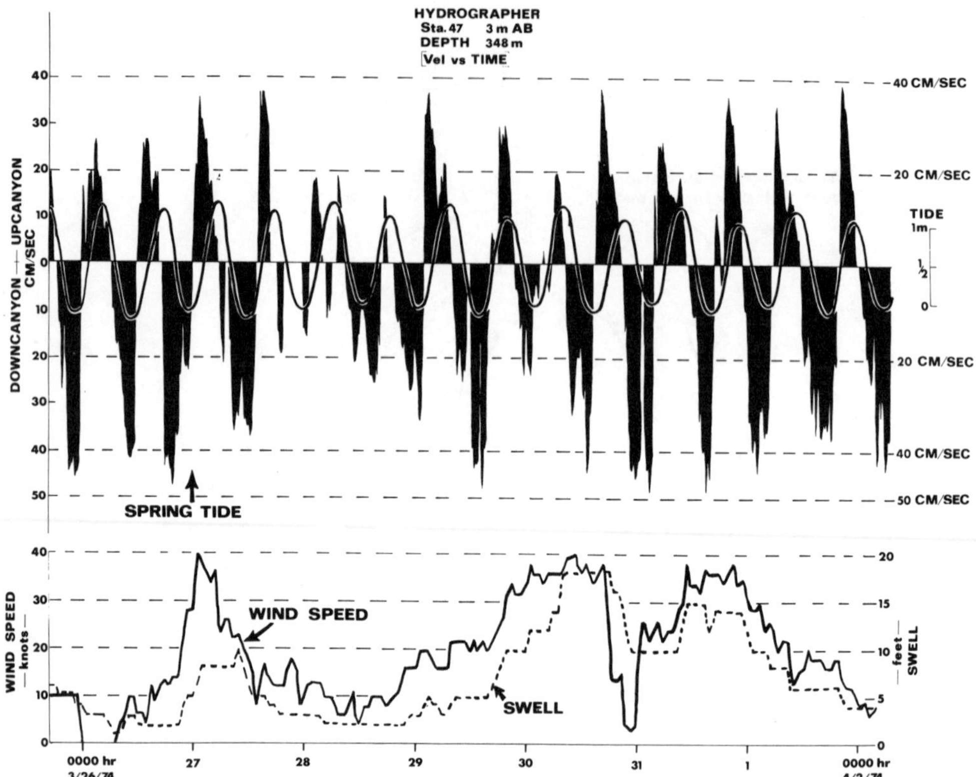

FIG. 85—Showing possible relation between storm and calm conditions to speed of currents at the shallow station in Hydrographer Canyon. The small currents during the calm on May 31 may be related to other causes, as two of the fastest currents of the entire period occurred about March 26 and 27 during another calm period.

Possible relation to storm—During the period in which we had current meters in Hydrographer Canyon, we had one major storm with winds up to 50 knots (90 km/hour) and high swell conditions. The winds during the storm changed from northeast to west with a short period of calm in between. Plotted on the same graph as the time-velocity curve for the currents at the shoalest station, the wind and swell, as observed on the NOAA ship from which the work was accomplished (Fig. 85), show a possible relationship to this storm because the currents were weaker during the calm period that preceded. Even in this diagram we find that the highest speeds (downcanyon) occurred near the beginning of the first small storm and again near the beginning of the second storm and continued to some extent throughout the second storm. During the calm period between storms, the currents were slower, but this agreement may be pure coincidence, particularly because higher current speeds (also downcanyon) occurred near the beginning of the record when there was no storm.

Shortly after a hurricane passed over the New York Bight in 1971, turbid water was seen in Hudson Canyon during a dive to 273 m, where relatively clear water was observed both before and some time after the storm (Keller and Shepard, 1978).

This suggests that where a canyon head is well out from shore only storms of very high intensity initiate significant sediment transport.

Hudson Canyon

The head of Hudson Canyon is close to the outer terminus of the shallow shelf valley which extends into New York harbor. The canyon penetrates into the shelf for 30 km. Seaward the canyon follows a winding course with steep walls and many small tributaries (Fig. 86). It can be traced out to a depth of about 4,500 m, the outer part being a combination of fan valley and gorge. Most of the inner canyon has a mud bottom unlike the Georges Bank Hydrographer Canyon that is predominantly floored with sand and rock.

Extensive study was made of the currents in Hudson Canyon. Near the head, an "erratic and turbulent flow" was observed from a deep-diving vehicle (Keller and Shepard, 1978), and these flows occasionally were seen transporting sediment up to gravel size. A current meter was stationed in Hudson Canyon at 800 m for 49 days (A. Amos, personal commun.). It showed a net upcanyon transport. The highest velocities at 30 cm/sec were downcanyon, but most of the relatively high-velocity flows were upcanyon. On the other hand, current-meter records acquired during a submersible dive at about 3,000 m showed a continuous downcanyon flow at 7 and 100 m above the axis for four days (Cacchione et al, 1978). The speeds did not exceed 15 cm/sec. They were modulated by tides but never reversed.

During the Keller operations in 1974, current meters were placed in Hudson Canyon at depths of 223; 1,222; and 1,765 m. At the shallow station 3 m above bottom, a record was obtained from March 19 to 25, but at 30 m above bottom the record continued until April 4 (Fig. 87). At the two deeper stations the records were from April 13 to 26, both recorded at 3 m above the bottom (Fig. 88). As in Hydrographer Canyon, the fastest currents were found at the shallow station although they did not exceed 33 cm/sec which is definitely slower than the currents at the head of Hydrographer Canyon. The currents were weaker at the deep stations with maxima of 24 cm/sec at 1,222 m and 18 cm/sec at 1,765 m (Fig. 88).

In Hudson Canyon our progressive vector diagrams showed net downcanyon flow at the two shallower stations and upcanyon or crosscanyon at the deep station. The Hudson Canyon currents appear to flow almost as actively in an upcanyon direction as they do downcanyon.

The polar plots from the 1,222-m and 1,765-m records are somewhat puzzling (Fig. 89). At 1,222 m the up- and downcanyon flow directions show a very wide spread averaging northwest for upcanyon and southeast for down. At 1,765 m the direction seems to extend all the way from about 190° around to 80° with little concentration in any of these direction. The downcanyon direction of 70° was chosen by averaging the directions shown on the original record. This admittedly does not agree well with the supposed southeasterly direction of the axis; however, upon examination of the polar plot one realizes there is virtually no current in that general direction. It may be that the canyon floor is wide enough here so that the flows are not confined although profiles do not indicate a wide canyon.

According to the records, the cycles of up- and downcanyon alternations showed an average of 12 hours at the 223-m station, 13.1 hours at the 1,222-m station, and 12.4 hours at 1,768 m. Thus the longest cycles are at the intermediate depth. The

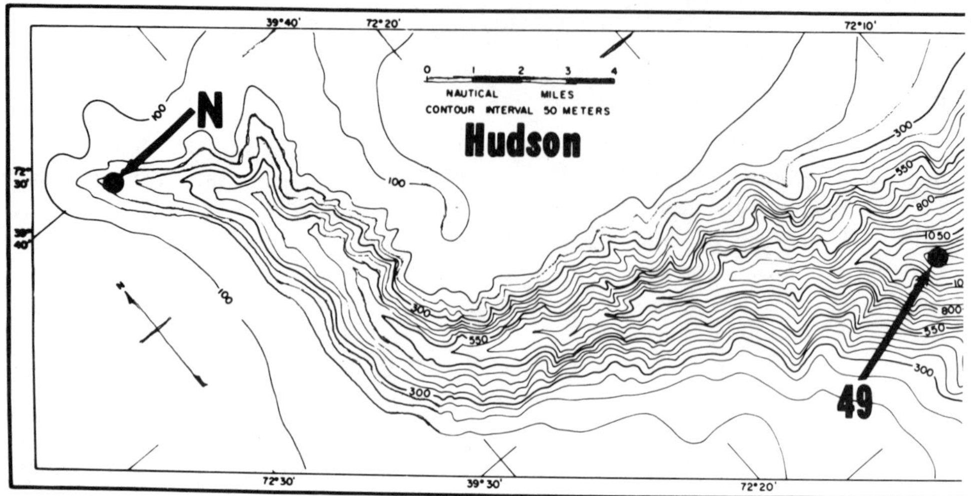

FIG. 86—Contour chart of Hudson Canyon with location of current-meter stations. *N* is Survey.

relation between the tide curves and the up- and downcanyon flows is clearly indicated in all of the time-velocity curves in Hudson Canyon. In comparing the time-velocity plots at 1,222 m and 1,765 m (Fig. 88), the deeper station shows a change so that the upcanyon peaks start during low tides and later shift to correspond to high tides, whereas they remain quite consistently close to low tides at the 1,222-m station. This shift in the 1,764-m station is not continuous; the last half of the record shows most of the upcanyon peaks close to the time of high tide. It is hard to tell what significant relationship there is in this shift in relation to the tidal phase.

Wilmington, Washington, and Norfolk Canyons

Off the Mid-Atlantic States there are several canyons, all somewhat smaller than Hudson and Hydrographer. A few current-meter stations were obtained by George Keller in Wilmington, Washington, and Norfolk Canyons. Also stations were located on the shelf east of Wilmington Canyon and on the slope east of Norfolk Canyon.

Wilmington Canyon, located off Delaware Bay, is cut into the continental shelf for a distance of about 8 km. It has a winding course (Fig. 90). This canyon has been sampled extensively by Stanley (1974) and Stanley and Kelling (1967). Short-period current observations were made by Stanley et al (1972).

The investigation by Keller involved the use of one of our current meters at 914 m and a NOAA meter at 725 m. Both of these agree almost perfectly in their up- and downcanyon alternation cycles with the tides (Fig. 91). Because they were not taken at the same time, it is impossible to look for internal wave advance. Comparison of curves for 3 and 30 m above the bottom at the shallow station (Fig. 91B) shows some local agreement but elsewhere the flow of the two are in opposite directions. The progressive-vector diagrams show that the net flow at the shallow station is upcanyon at 3 m above bottom and downcanyon at 30 m. At the deeper

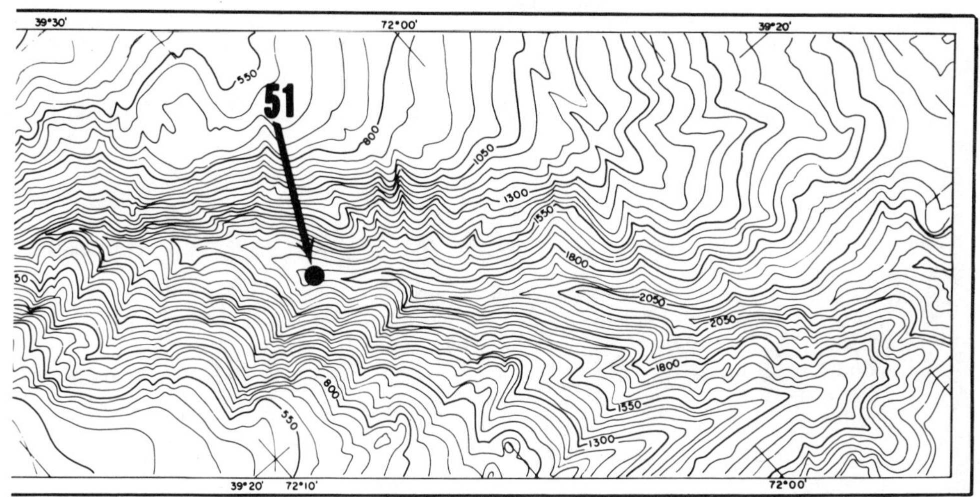

NOAA current meter. Contours by K. O. Emery from soundings by U.S. Coast and Geodetic

station the progressive-vector diagram shows that the net flow is variable but slightly upcanyon (Fig. 92B). The polar plot at 914 m indicates rather broad dispersal of flow direction but with a general grouping northwest for upcanyon and southeast for downcanyon (Fig. 92A).

The current velocities are higher than at the shoaler station with maximum velocities of about 40 cm/sec at 725 m and about 21 cm/sec at the 914-m station.

Despite the indication that the net water movement is upcanyon, there are indications that the sediment transport may be downcanyon. The discovery of ripple marks with steep slopes downcanyon and the nature of material on the canyon floor suggest major downcanyon transport (Stanley et al, 1972).

A single current meter was placed in Washington Canyon. This canyon cuts into the shelf for about 13 km. It becomes a fan valley with well-developed levees at about 800-m axial depth (Fig. 93). It appears to have a less sinuous trend than other East Coast canyons and is less incised into the slope. The bottom photographs in this canyon show ripple marks out to 740 m (Keller and Shepard, 1978) and "shell hash," suggesting active flows occur, although the floor becomes muddy at a depth somewhat over 700 m.

A 15.5-day record was obtained with a NOAA current meter. The flows did not exceed 23 cm/sec and were primarily up and down the axis. Unlike Wilmington Canyon, the net flow in Washington Canyon is downcanyon. Velocities were low, averaging only 6 cm/sec. The record from the nearby slope also showed low-velocity currents with a dominant southeasterly direction (Keller and Shepard, 1978).

Norfolk Canyon, located approximately off Chesapeake Bay, indents the continental shelf about 16 km and is cut down about 800 m into the slope, much deeper than neighboring Washington Canyon (Fig. 93). Historically, drainage into Chesapeake Bay (including the Susquehanna River) was flowing into the head of this canyon during the lowered sea levels of the glacial stages.

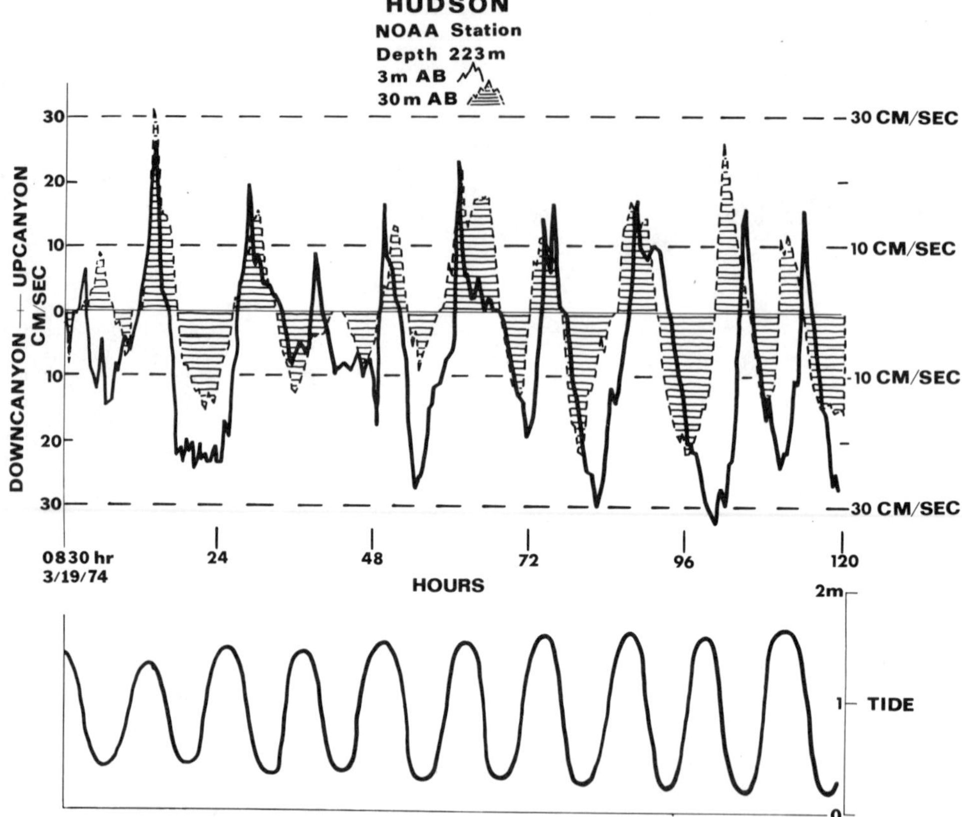

FIG. 87—Overlay of the time-velocity curves from records at 3 and 30 m above bottom in 223-m axial depth of Hudson Canyon. Supplied by George Keller from NOAA records.

NOAA current meters were placed at 3 m and 30 m above bottom at an axial depth of 575 m although only the one at 30 m above the bottom worked. This 15-day record shows velocities up to 30 cm/sec both for up- and downcanyon flows. It shows the usual semidiurnal tidal alternation of up- and downdanyon with net downcanyon flow. In the progressive-vector diagram a strong easterly drift is shown which is not unusual at 30-m height above bottom in East Coast canyons.

West Indies Canyons

The submarine canyons around the West Indies are for the most part poorly sounded. An exception is found off Puerto Rico and along the north slope of St. Croix Island. Some of these canyons are associated with terrestrial drainage. The heads of these canyons extending in close to the shore, like those off California, have steep gradients and rather precipitous walls. Some of them appear to terminate part way down the submarine slopes of the islands.

Rio de la Plata Canyon

The U.S. Geological Survey investigated some of the canyons off northern Puerto Rico. J. V. Trumbull of the Survey arranged for us to place two current meters into

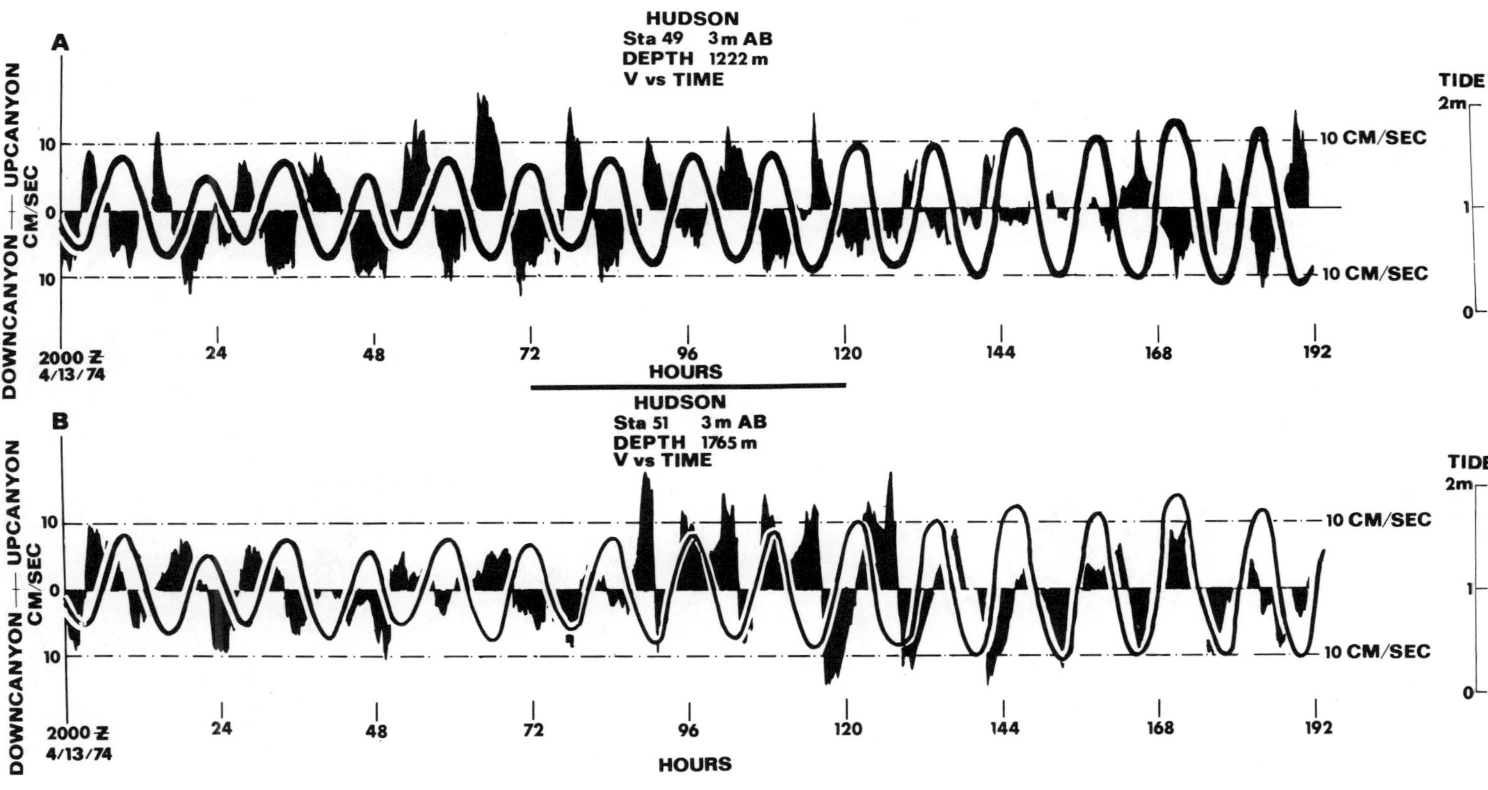

FIG. 88—Time-velocity curves of part of the record at 1,222- and 1,765-m depths in Hudson Canyon. There is a rather good agreement of the currents with the tides for the entire 1,222-m station, but the peak flows at the 1,765-m station change their relation to the tides as the record continues.

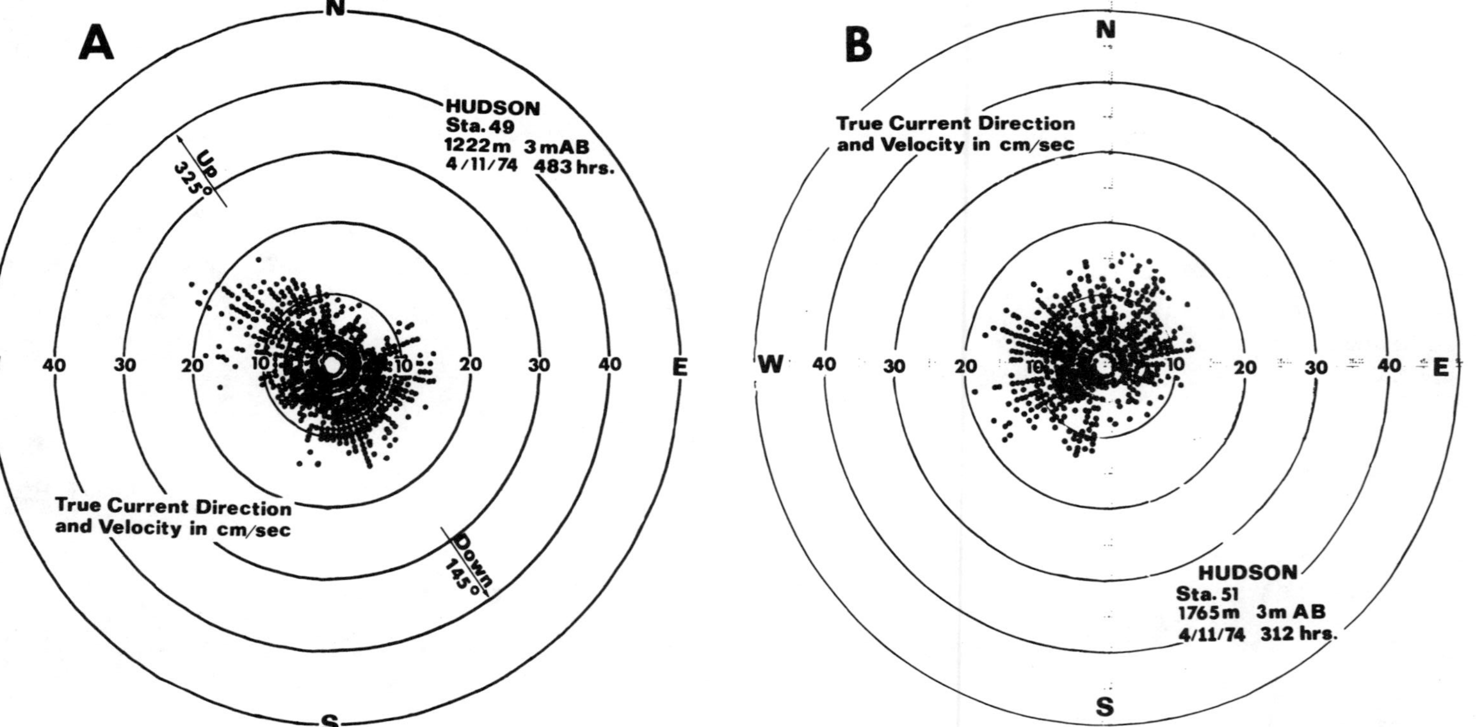

FIG. 89—Polar plots of records from 1,222 and 1,765-m depths in Hudson Canyon. At the shoaler station some relation of current flow to the canyon axis is indicated but none is apparent at the deeper station. Note how slow these currents were, especially at the deep station.

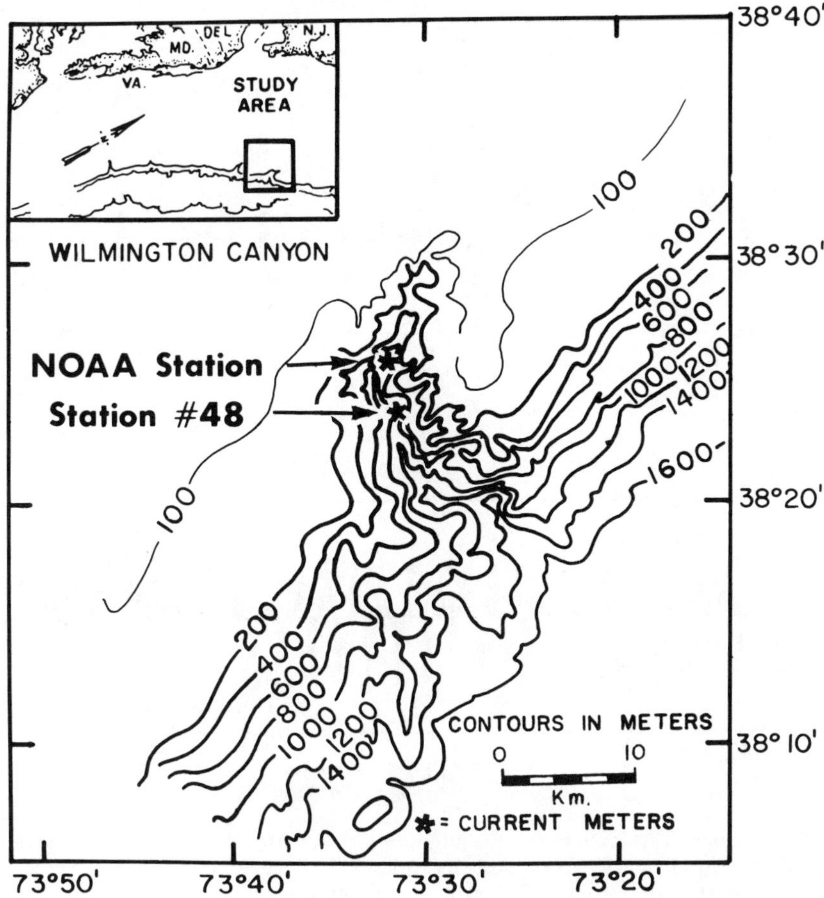

FIG. 90—Contour chart of Wilmington Canyon, showing location of current-meter stations. Contours from U.S. Coast and Geodetic Survey soundings. Inset shows location relation to coast.

Rio de la Plata Canyon. This canyon heads (at about 50-m depth) a kilometer from the coast directly off a river of the same name (Fig. 94), and has been studied by Pilkey et al (1978). Other canyons along this coast appear to come in even closer. A short distance offshore, Rio de la Plata Canyon enters a ravine cut from 200 to 300 m below the surrounding slope. So far as known, the canyon becomes much less incised at greater depths (the Rio de la Plata was in flood stage when the current meters were emplaced).

One meter was deployed in the axis of the canyon 1.7 km from shore at a depth of 238 m and was left there for 84 hours. The other current meter was emplaced att 439 m, 3.5 km from shore, and was left there for 649 hours. Both meters were suspended 3 m above the bottom. The tides in the area are very small with a range of about 0.4 m.

The instrument at 238-m depth showed remarkably fast currents both up- and downcanyon (Fig. 26). The currents are somewhat faster downcanyon with speeds

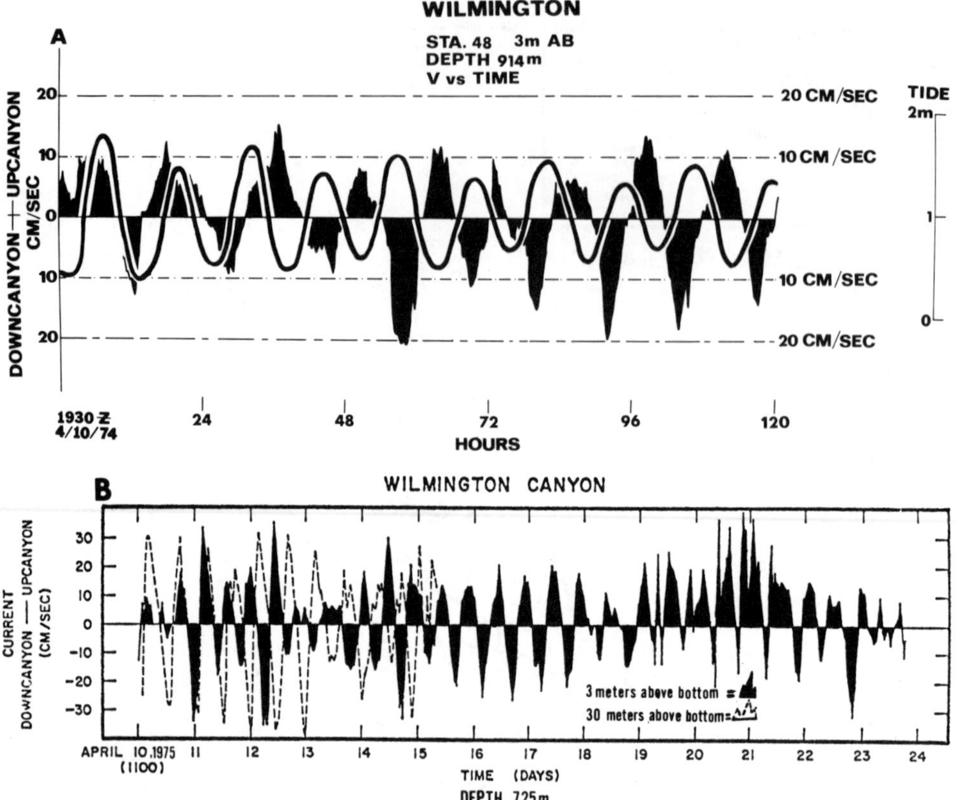

FIG. 91—**A**, time-velocity curve for Wilmington Canyon at 914-m depth, 3 m above bottom. Note good relation of cycles to semidiurnal tide. NOAA current meters. **B**, overlay of time-velocity curves at 3 and 30 m at 725 m in Wilmington Canyon from a record obtained by George Keller with NOAA current meters. The currents of 40 cm/sec for both up- and downcanyon currents at both 3 and 30 m indicate that these currents are unusually fast, especially for the 30 m above bottom.

up to 50 cm/sec upcanyon and 48 cm/sec downcanyon (not including the 100 cm/sec determined for the turbidity current). Most of the currents flowed along or close to their axial direction (Fig. 95A).

The frequency of alternation of current flow is very rapid as is typcial of a small tidal range. The average period is only 2.1 hours. No agreement with the tide phases could be detected in the record, and the alternation-cycle lengths are irregular.

The record at 439 m (Fig. 96) is quite different from that at 238 m. The up- and downcanyon reversals are more irregular but still have a relatively high frequency, averaging about 3.5 hours or even shorter if one counts all of the very small alternations. This high frequency at such a considerable depth is again related to the small tidal range. The velocity is slightly lower but reaches 34 cm/sec downcanyon and 27 cm/sec upcanyon as seen on the polar plot, which also shows some spread of direction but considerable concentration in the upcanyon and downcanyon sectors (Fig. 95B). The net flow is downcanyon with fairly consistent direction.

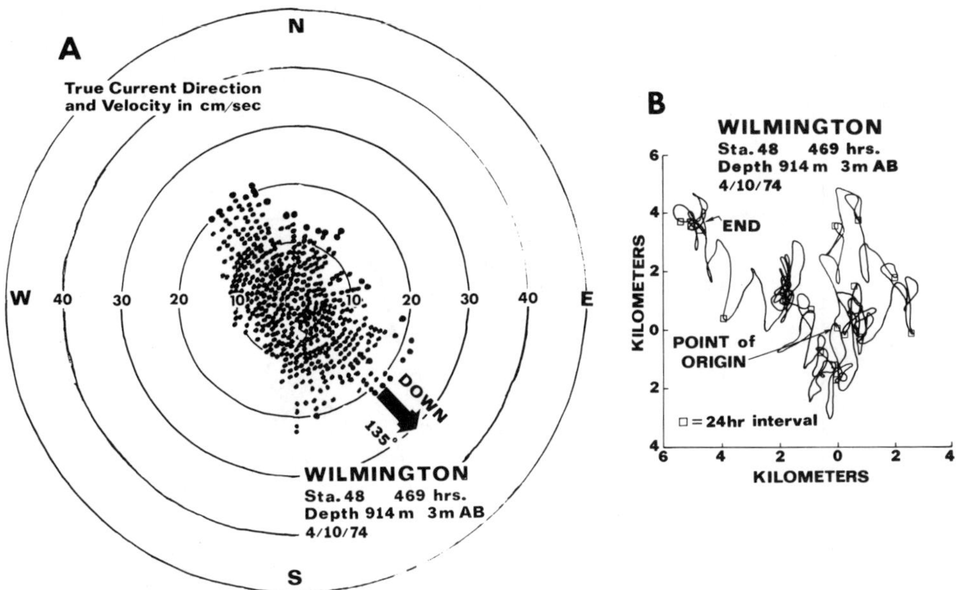

FIG. 92—**A**, polar plot for Wilmington Canyon at 914-m depth showing better agreement with axial trends of chart than in Hudson and Hydrographer Canyons. **B**, progressive vector diagram of same.

Turbidity Currents—At the end of the 238-m record we have what appears to be a turbidity current (Figs. 26, 27). We probably would not have recognized this as such had we not seen the result of a similar event in La Jolla Canyon (Fig. 20). There, as in this record, the downcanyon flow was preceded by a relatively strong upcanyon flow. The turbidity current here attained a speed of 100 cm/sec. Unfortunately, the record came to an end shortly after the strong surge while the current was returning to normal, so that we do not know whether or not there was the usual period of no current after the surge was over.

The probable explanation for a turbidity current here is that the river was carrying a large amount of sediment into the ocean at this time due to the floods. In fact the river gauge showed flood conditions peaking two days prior to the turbidity current (Fig. 97).

Evidence for the continuation of the turbidity current at the deeper station was not found. The first downcanyon flow as high as 30 cm/sec occurred four days after the turbidity current at the shallow station, and furthermore this very slight surge shows no indication of being a turbidity current. No large upcanyon flow precedes it, nor is there any period without a measurable current after this moderate surge (Fig. 98).

Salt River Canyon, St. Croix

On the north side of St. Croix in the Virgin Islands, a small submarine canyon heads directly off a gap in a coral reef (Fig. 99A). This canyon is well sounded only near shore and extends seaward for only 2 km before terminating on the slope. Inside the reef there is an estuary into which Salt River empties. In its lower course

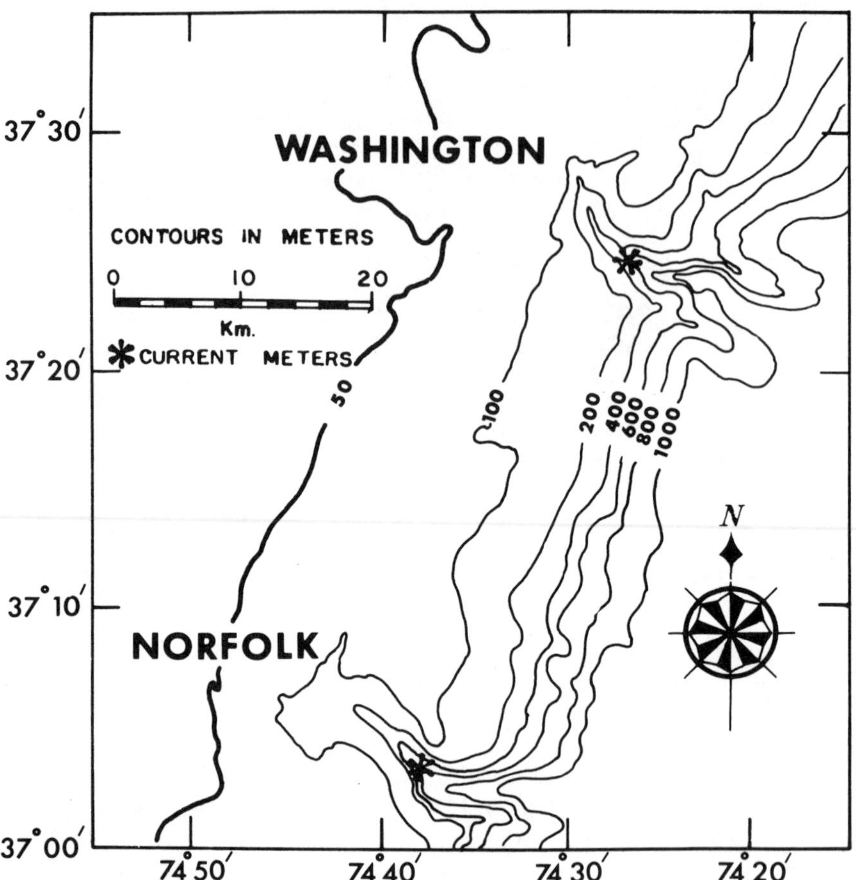

FIG. 93—Contour chart of Washington and Norfolk Canyons. Norfolk Canyon located off Chesapeake Bay has deeper penetration into the shelf than Washington Canyon.

the river flows through mangroves but has a definite incised valley further inland. The very small river flow during much of the year and high evaporation rate causes high salinities in the estuary except in the rainy season; hence the name. The canyon head has a very steep gradient so that a depth of 190 m in the canyon is only 0.5 km from the coral reef.

Through the cooperation of R. F. Dill and the West Indies Laboratory and later through an invitation from Orrin Pilkey to join an expedition of the Duke Univesity ship *Eastward*, we had the opportunity to work in this area. Our first operations on the West Indies Laboratory boat *Sarima* were conducted in June of 1976 during the dry season with rather light trade winds, and the second operation was in February of 1977, following a rainy period and during intermittently strong trade winds.

In the first season we obtained records in axial depths of 48, 90, and 164 m although the record at 90 m was of questionable value and at 164 m the 3-m above bottom record showed only directions. At 48 m the currents at 3 m above bottom were mostly very slow, less than 10 cm/sec, but with occasional downcanyon flows up to 30 cm/sec (Fig. 100). These relatively fast flows generally occurred during a

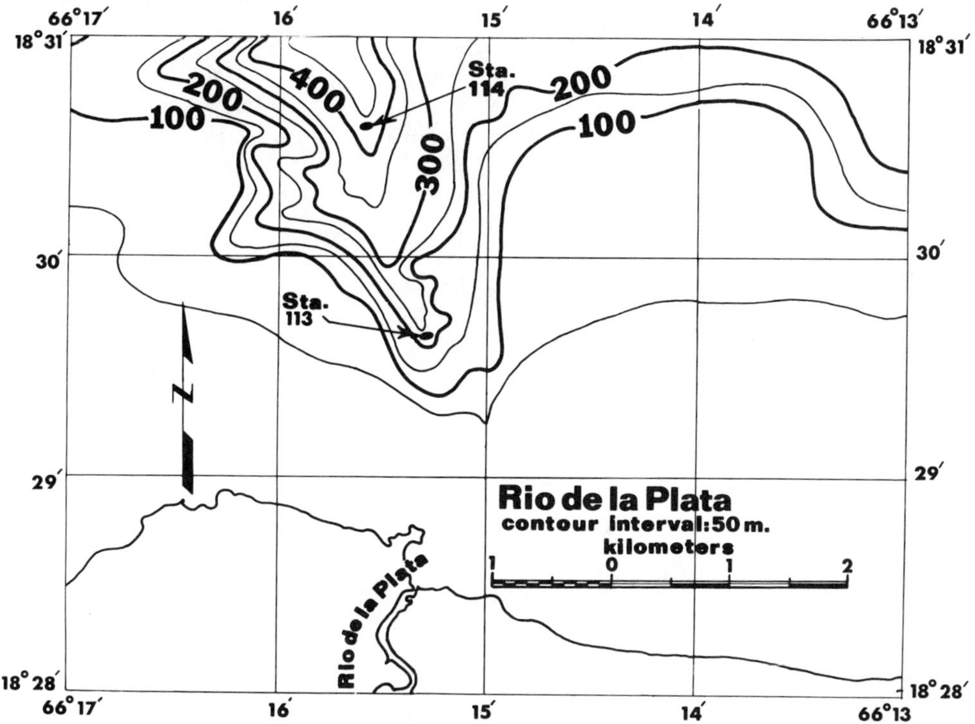

FIG. 94—Contour chart of Rio de la Plata Canyon showing relation to the river which was in flood during current-meter recordings. Position of current meters indicated. Contours from U.S. Geological Survey.

falling tide. The frequency of the changing directions is best shown by duplicating a part of the tape (Fig. 101). Here again we find frequent oscillation of direction coincident with a small tidal range (0.3 m or less).

The relatively strong downcanyon flows during some ebb tides are rather easily explained (Shepard and Dill, 1977). The high-density, salty water in the lagoon is probably carried out while the tide is falling, and flows down the head of the canyon. Our record at the shallow station 30 m above the bottom did not show this downcanyon flow, indicating it occurred only along the bottom. A rather strong downcanyon flow along the bottom near the same place was observed by R. F. Dill during a scuba dive. We attempted to check the explanation in returning to the area in February, 1977. Water temperature recorders were deployed by Neil Marshall and Gary Sullivan at various depths. These records indicate that the warm, saline water was flowing down the canyon head during ebbing tide. Probably the water was less saline in the bay at this time because of recent rains. This is supported by observations of mud clouds coming out of the estuary during ebbing tide. These moved west along the coast under the influence of the trade winds.

Our current-meter measurements at about the same depth (49 m) in February of 1977 produced very different results than those of June, 1976. At 3 m above the bottom the 5 days of the record showed an almost constant downcanyon current with the exception of a few very short upcanyon flows (Fig. 10B). This unidirection-

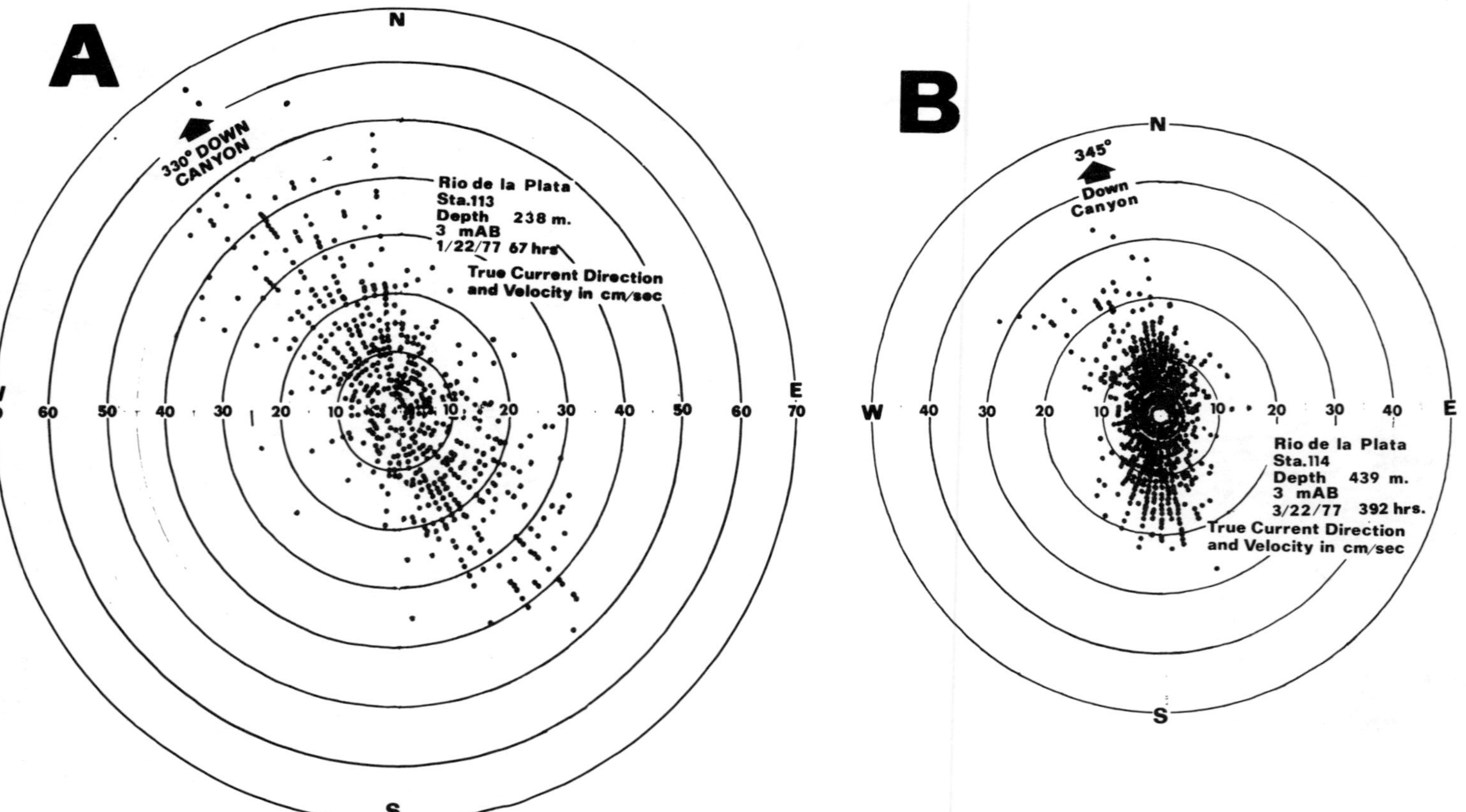

FIG. 95—Polar plots of the currents at the two stations in Rio de la Plata Canyon. The fastest currents at Station 113 represent a turbidity current.

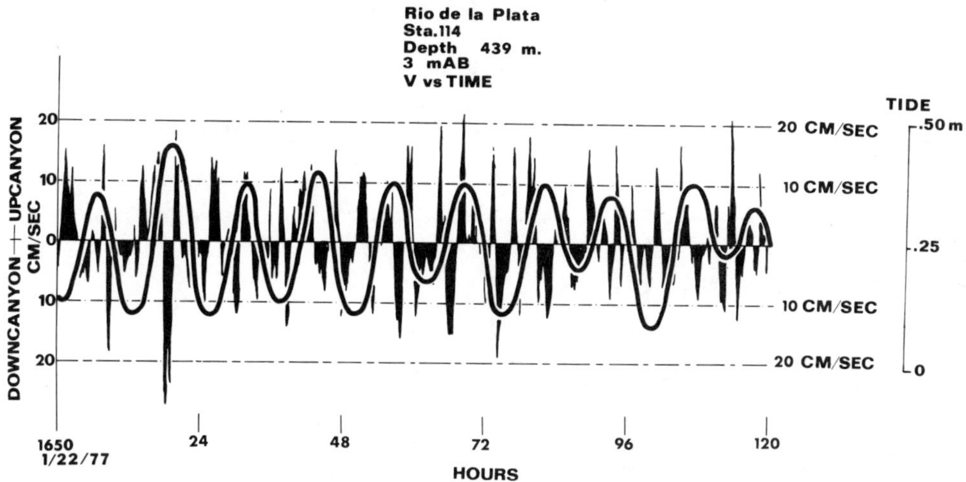

FIG. 96—Portion of time-velocity curve in Rio de la Plata Canyon at a depth of 439 m. Note the considerably lower speeds from those observed at 238 m (Fig. 26) and the absence of any sign that the turbidity current continued out to that depth (Fig. 98).

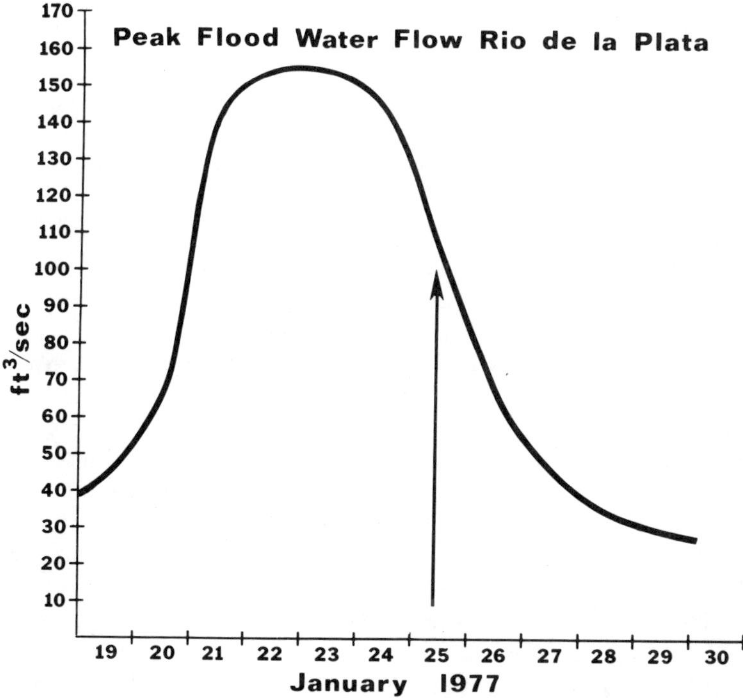

FIG. 97—Figure showing time of the peak of the flood waters coming down the Rio de la Plata and comparison of this peak with time of the turbidity current (arrow) in Rio de la Plata Canyon. Clearly the sediment introduced by the flood led to the subsequent turbidity current.

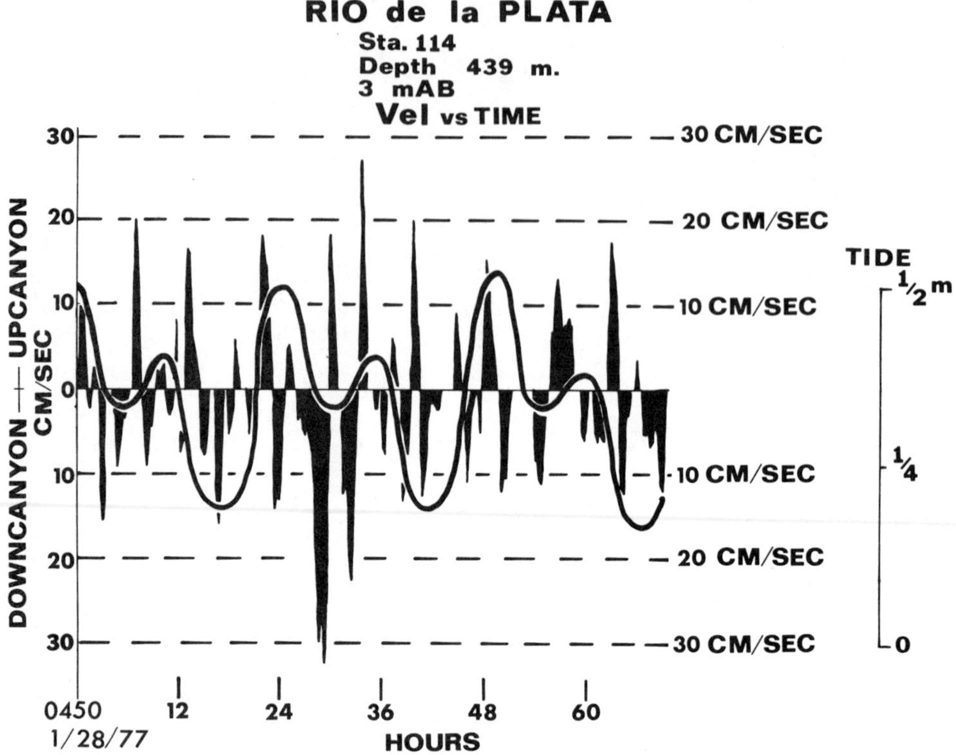

FIG. 98—Portion of time-velocity curve at 439-m depth in Rio de la Plata Canyon showing the first relatively fast downcanyon flow after the time of the turbidity current at the 238-m station. This occurred too late to be a downcanyon continuation of the turbidity current.

al flow is unique in our records. On the other hand, at 30 m above the bottom the current was flowing upcanyon most of the time (Fig. 10A). Comparing the records at the two levels we see a frequent relationship between the rare downcanyon flows at 30 m and the relatively fast downcanyon surges in the 30-m record. An exception exists at hour 96 where the downcanyon flow at 30 m did not influence that at 3 m.

The isothermal water conditions and the strong trade winds probably are responsible for the predominant upcanyon flow at 30 m and for the almost continuous downcanyon flow at 3 m. At 30 m above the bottom the trade wind-induced currents apparently are effective 19 m below the surface, driving the water shoreward in a southerly direction, as indicated by the polar plot (Fig. 102). As a result there had to be a return flow, and because conditions were isothermal the water could return along the bottom at least in the head of the canyon. Possibly the occasional rather strong downcanyon surges at 3 m extended upward to the 30-m level, developing an exceptional seaward flow through most of the water column at these times.

The polar plot at 3 m above the bottom shows very little flow in the upcanyon direction. Virtually all the data points showing currents of over 10 cm/sec are in the downcanyon sector. A few flows in the SSW direction suggest temporary influence of the surface wind current.

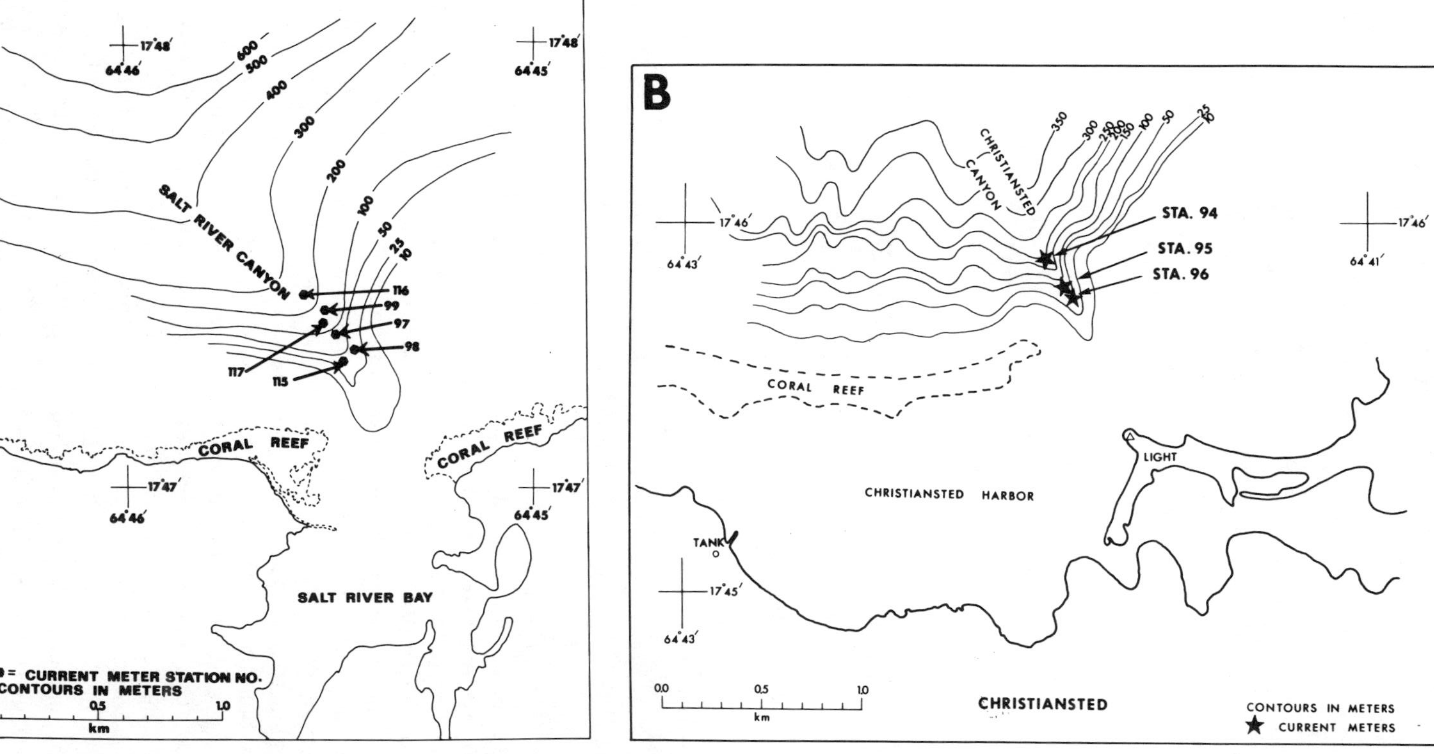

FIG. 99—**A**, Salt River Canyon based on surveys by R. F. Dill. Shows current meter stations. Note relation to gap in coral reef and lagoon. **B**, head of Christiansted Canyon showing relation to harbor entrance and location of current-meter stations. From surveys by U.S. Coast and Geodetic Survey and by the writers.

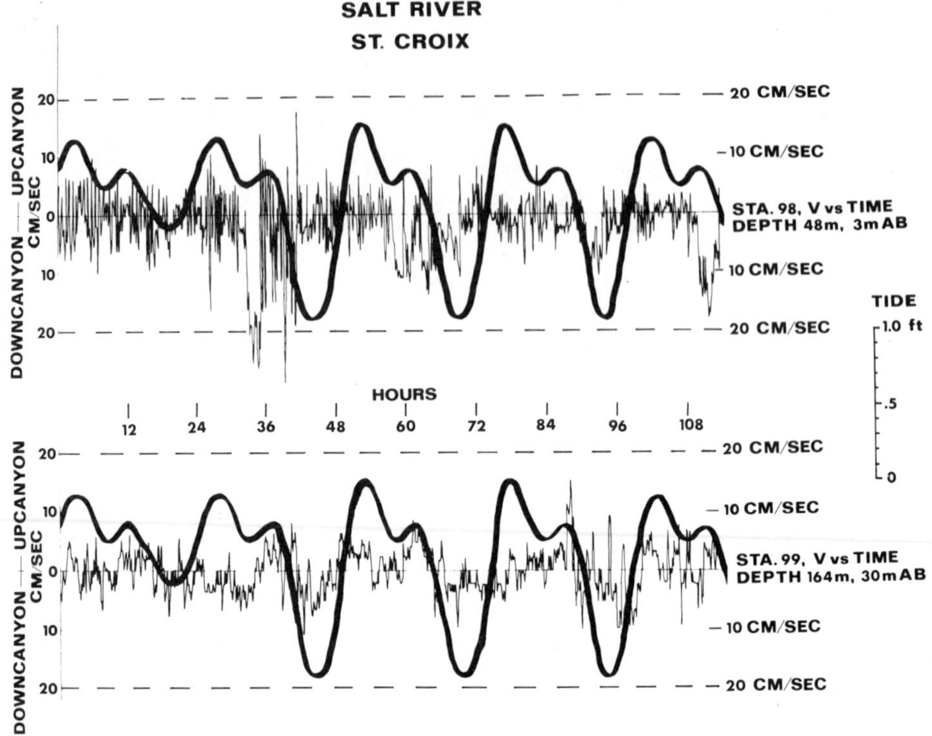

FIG. 100—Time-velocity curves in Salt River Canyon at 48 and 164-m depths. Notice the high frequency of direction alternation in this area of small tidal range. The somewhat faster downcanyon flows during some of the ebbing tides may be the result of high salinity water coming out of the estuary and flowing down the canyon head.

In February, 1977, additional stations were located at depths of 130 and 238 m. Both had current meters at 3 and 30 m above the bottom. At the 130-m station the record from 3 m showed irregular currents with more alternations in speed than in direction (Fig. 103A). Using only the first 60 hours of data (because there were several gaps after that), we found an average cycle of direction alternation of 2.5 hours. The net flow was slightly downcanyon. The velocities are all slow with a maximum of 11 cm/sec. At 30 m above the bottom the currents were slightly faster with a maximum of 13 cm/sec. There are long periods of continuous downcanyon flow, particularly during the first 18 hours when it was almost continuously in that direction, but later in the record, upcanyon flows also continued for long periods; however, net flow was downcanyon. The alternation cycles average about 4.5 hours, although this average may be somewhat meaningless in this case.

At the 238-m station (Fig. 103B), the 3 m above bottom current had unusually frequent alternations in direction of flow for such deep water, approximately 198 cycles in 210 hours with hardly any appreciable periods of unidirectional flow. The currents never exceeded 6 cm/sec and were almost well under 5 cm/sec. Net flow was upcanyon at 3 m above bottom. At the same station at 30 m above the bottom, the alternations were almost as short. Directions of flow showed little relation to the canyon axis, and the two highest velocities were 9 cm/sec, one east and the other west.

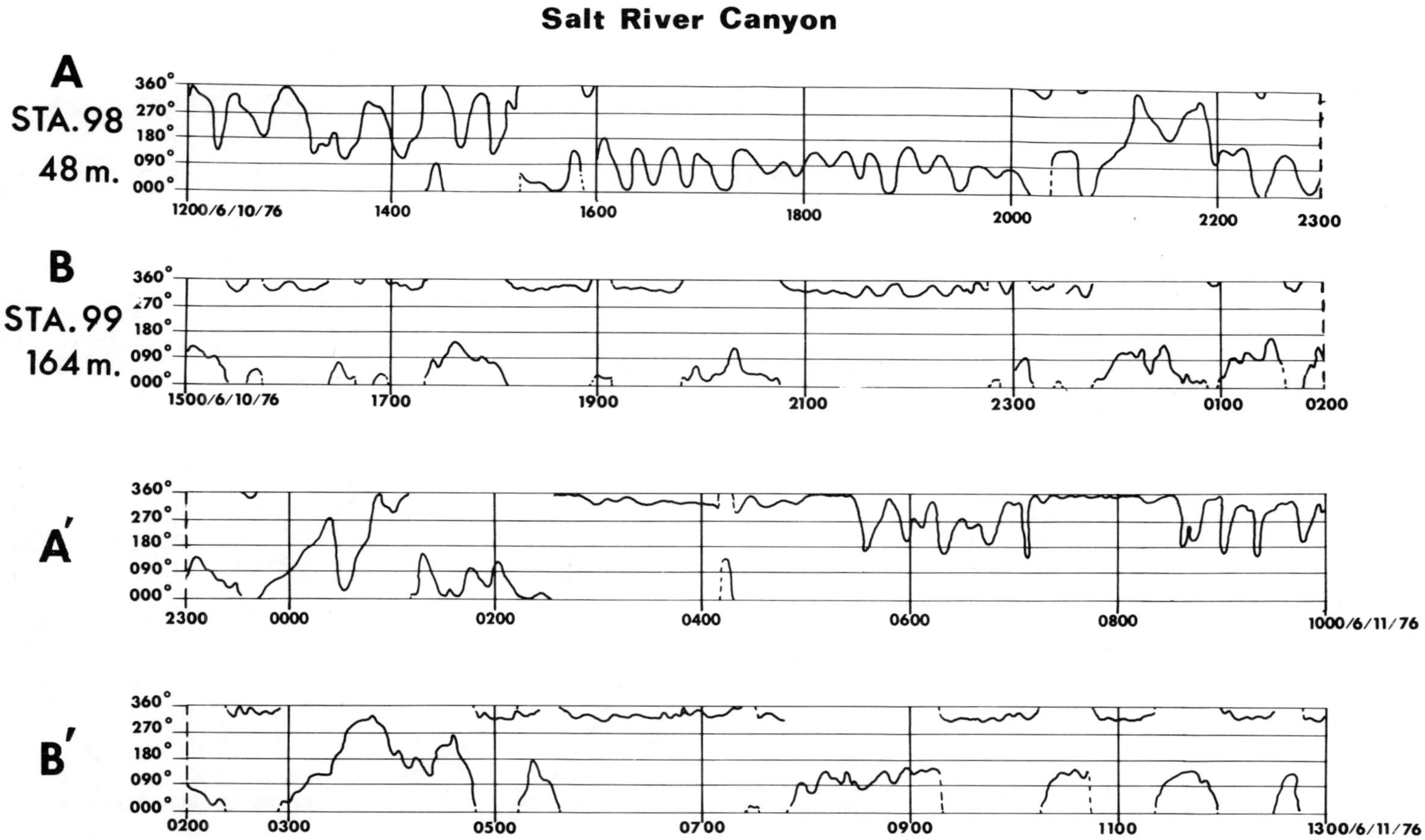

FIG. 101—Copy of the original current-meter record showing the frequent alternations of flow in Salt River Canyon that are not adequately indicated in time-velocity plots. The relatively long periods with constant direction of flow were probably due to ebbing tide carrying high-salinity water out from the estuary.

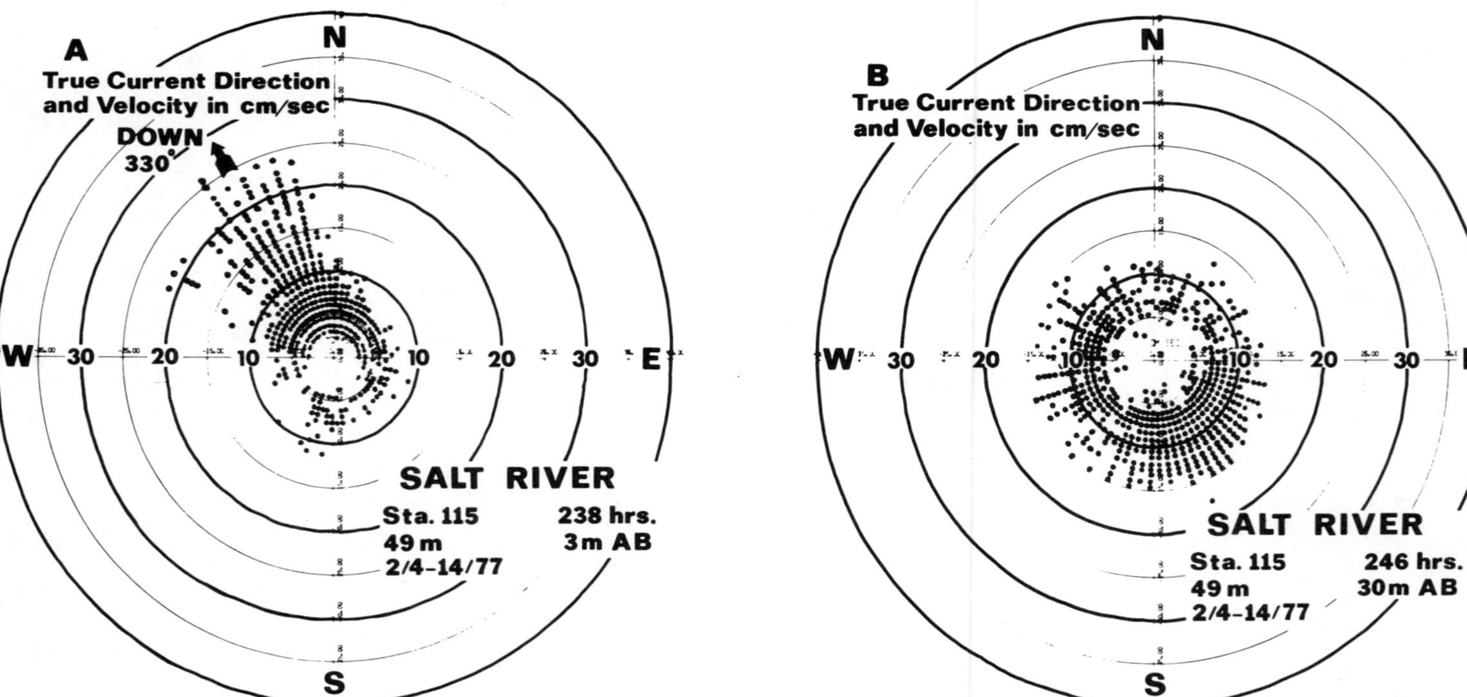

FIG. 102—Polar plots of the currents shown in Figure 10. Note the scarcity of flows of very low velocity at both 3 m and 30 m above bottom.

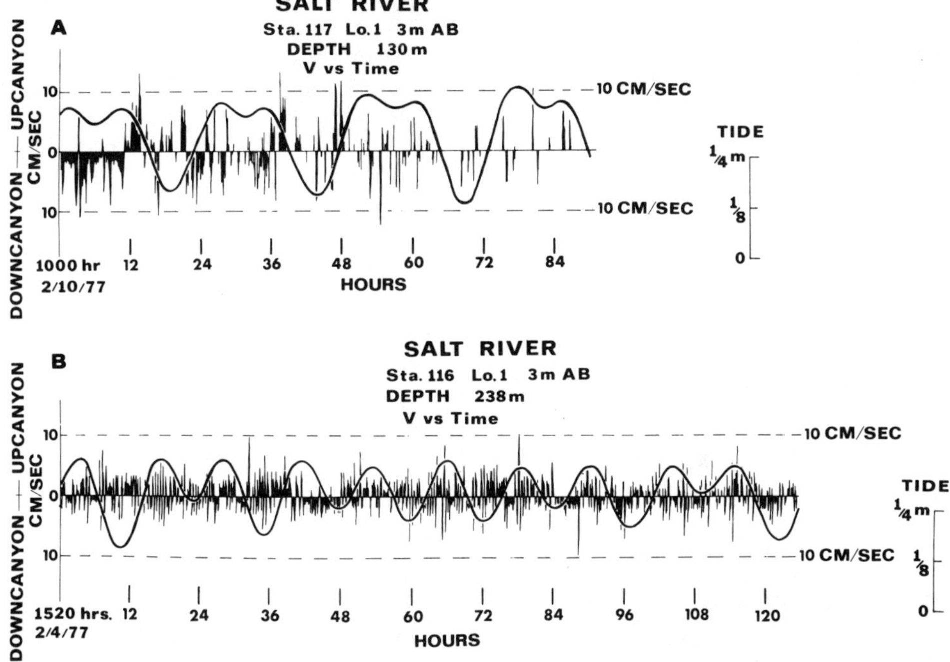

FIG. 103—Time-velocity curves at 130 and 238-m depths in Salt River Canyon taken contemporaneously with the records in Figure 10 shows that the constant downcanyon flows at 49 m did not continue to these greater depths. Frequency of direction changes was higher in this case at the station with greater depth.

Thus the currents in the shallow parts of Salt River Canyon are very unusual compared to other canyons. Alternation cycles are almost entirely missing in one of the records, and instead there are long, random-length periods of either up- or downcanyon flow.

Christiansted Canyon, St. Croix

A much larger canyon than that off Salt River is located off the break in the coral reef where the channel extends into Christiansted (Figs. 99B, 104). The canyon has a steep gradient and continues out about 9 km to where it apparently terminates at about 2,600 m on the slope leading down into the Virgin Island Basin. One cross section shows relief of about 1,200 m. The trend is generally to the north and is somewhat winding. Locally there are steep slopes on the sides of the canyon. So far as can be told from the rather inadequate survey, the only tributaries are at the head of the canyon. This is certainly quite different from the canyons with their many tributaries that are found off east and west coasts of the United States.

The canyon appears to be cut into a fault scarp off the north coast of St. Croix, and the slope continues to at least a depth of 3,900 m, where it appears to be overlapped by sediments of the basin.

In June of 1976 we obtained records at two shallow stations in Christiansted Canyon and in February of 1977 at four deep stations, extending the work out to a depth of 2,525 m near the terminus of the canyon.

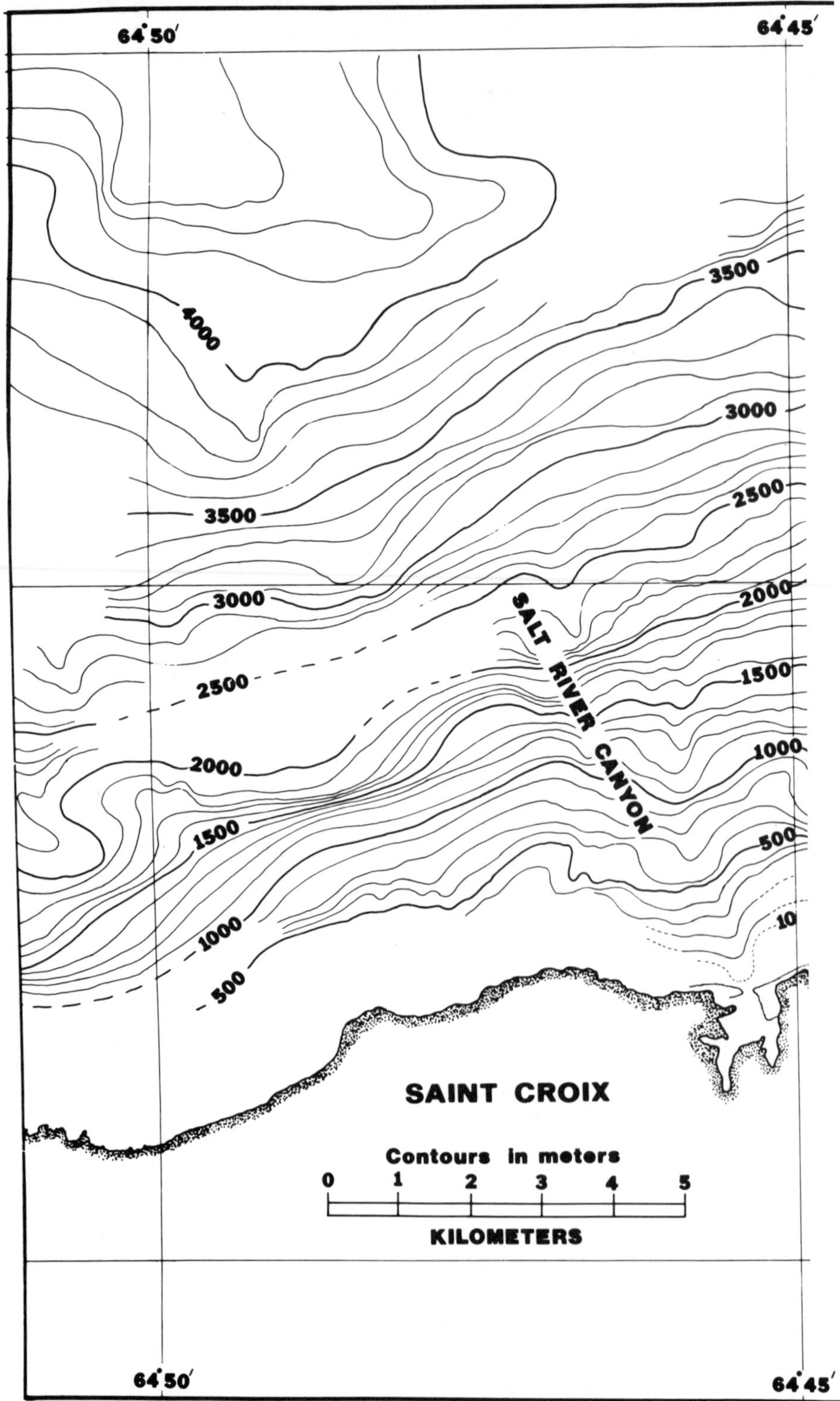

FIG. 104—Contour map of the Christiansted Canyon that extends down the escarpment on the north side of St. Croix Island. Current-meter stations are indicated except for those at the canyon head shown in Figure 99B. Considerable difficulty was found in fitting the sounding

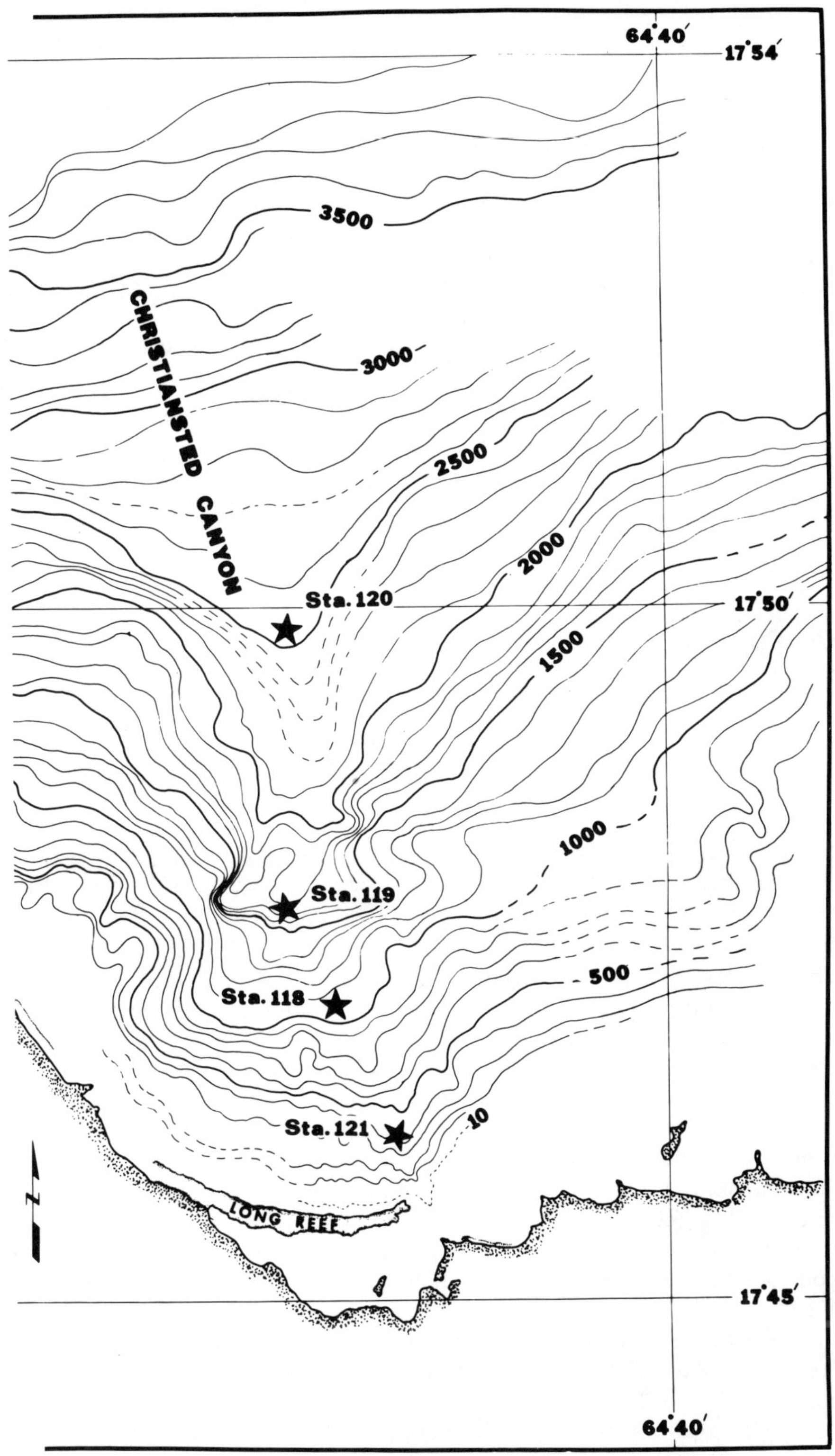

lines together because of somewhat inadequate survey methods for positioning. The writers were considerably helped in adjusting the lines by Tom Harper. The survey which was run on the Duke University *R. V. Eastward* was under the direction of R. F. Dill and F. P. Shepard.

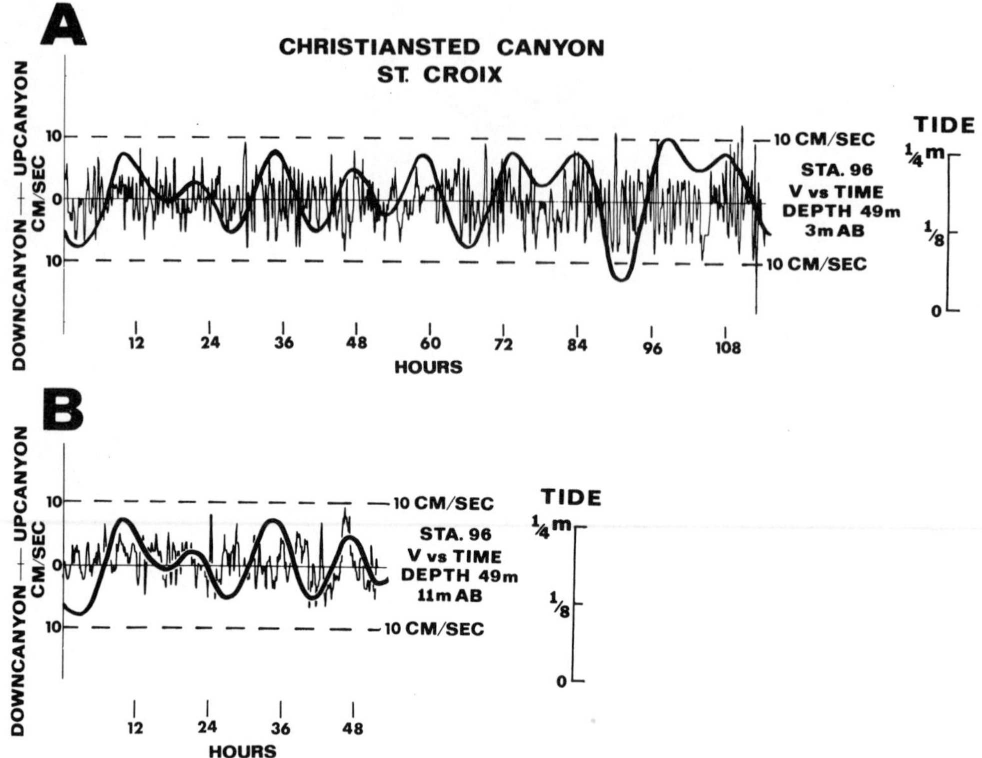

FIG. 105—**A**, time-velocity curve from 49 m at 3 m above bottom in the head of Christiansted Canyon. Another example of high-frequency direction alternations with small tides. **B**, time-velocity curve at 49 m, 11 m above bottom.

The shallow stations at 49 m with current meters at 3 and 11 m above bottom and at 56 m with current meters at 3 and 30 m above bottom were both examples of the high frequency direction alternations characteristic of small tidal ranges and shallow water (Fig. 105). The periods appear to be close to one hour at 56 m but two hours at the 49-m station. The currents were very slow, less than 10 cm/sec at the 49-m station, but attained velocities up to 20 cm/sec at the 56-m station. The net flow of the shallow station was downcanyon 3 m above bottom and upcanyon 11 m above bottom; similar to the 49-m station in Salt River Canyon, however, it lacked unidirectional flow (compare Fig. 10 and Fig. 105). At the deeper station, 30 m above bottom, the maximum speed was 20 cm/sec downcanyon and 11 cm/sec upcanyon, but the net flow was almost balanced with a very slight downcanyon preference.

Additional stations were located in the outer part of Christiansted Canyon at axial depths of 1,050; 1,765; and 2,525 m. One other station at about 410 m was in a tributary canyon to the west of the main canyon. At the latter we found weak currents at 3 m above bottom, but the flow periods are distinctly longer than in shallow water although they all show alternations of periods of weak current flow with times of no measurable current (Fig. 106). The average period was somewhat over 5 hours, and at times the direction was unchanged for more than 12 hours. The

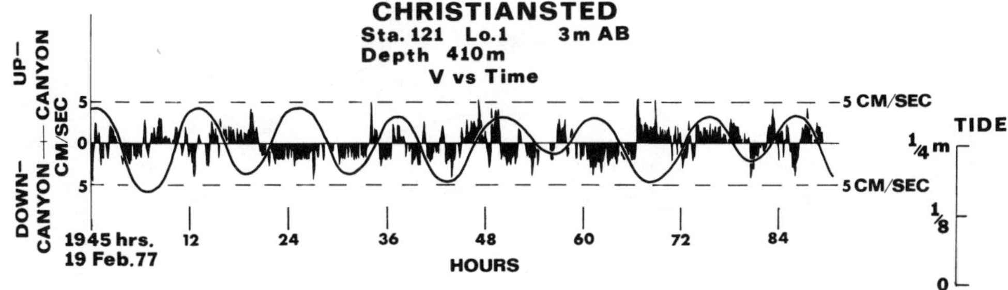

FIG. 106—Time-velocity curve at 410-m depth in tributary to Christiansted Canyon shows frequent alternations between weak currents and short periods of no recordable currents.

velocity in this record never exceeded 7.5 cm/sec. The up- and downcanyon flow balanced. Again there was little regularity in the periods of direction alternation.

At the 1,050-m station 3 m above bottom, regular alternating up- and downcanyon flows were again observed (Fig. 107). On the other hand, this record shows few of the alternations between 0 and a few cm/sec that characterize most of the shallower records in this St. Croix area. The average alternation cycle was 3.2 hours. The polar plot indicates that the directions are decidedly up- and downcanyon with very little crosscanyon (Fig. 108A). The crosscanyon time-velocity plot (not shown) confirms the absence of any fast currents across the axis. The maximum velocity was 15 cm/sec downcanyon and 11 cm/sec upcanyon. The progressive-vector diagram indicates a fairly consistent downcanyon net flow. At the same station 30 m above bottom somewhat slower currents were observed with maxima of 11 cm/sec downcanyon and 10.5 cm/sec upcanyon, and the alternation cycle averages 2.9 hours, a slightly higher frequency than at 3 m. The net flow at 30 m is upcanyon, contrary to what was found at 3 m above bottom.

At a depth of 1,765 m, the 3-m above bottom record again shows a rhythmic alternation of up- and downcanyon flows. The average cycle is 6.5 hours, showing

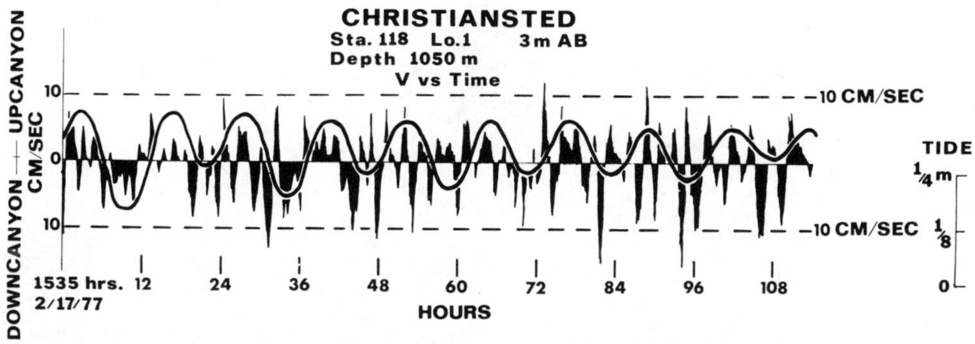

FIG. 107—Time-velocity curve from record in 1,050-m depth in Christiansted Canyon shows frequent reversal in direction of flow and relatively small currents.

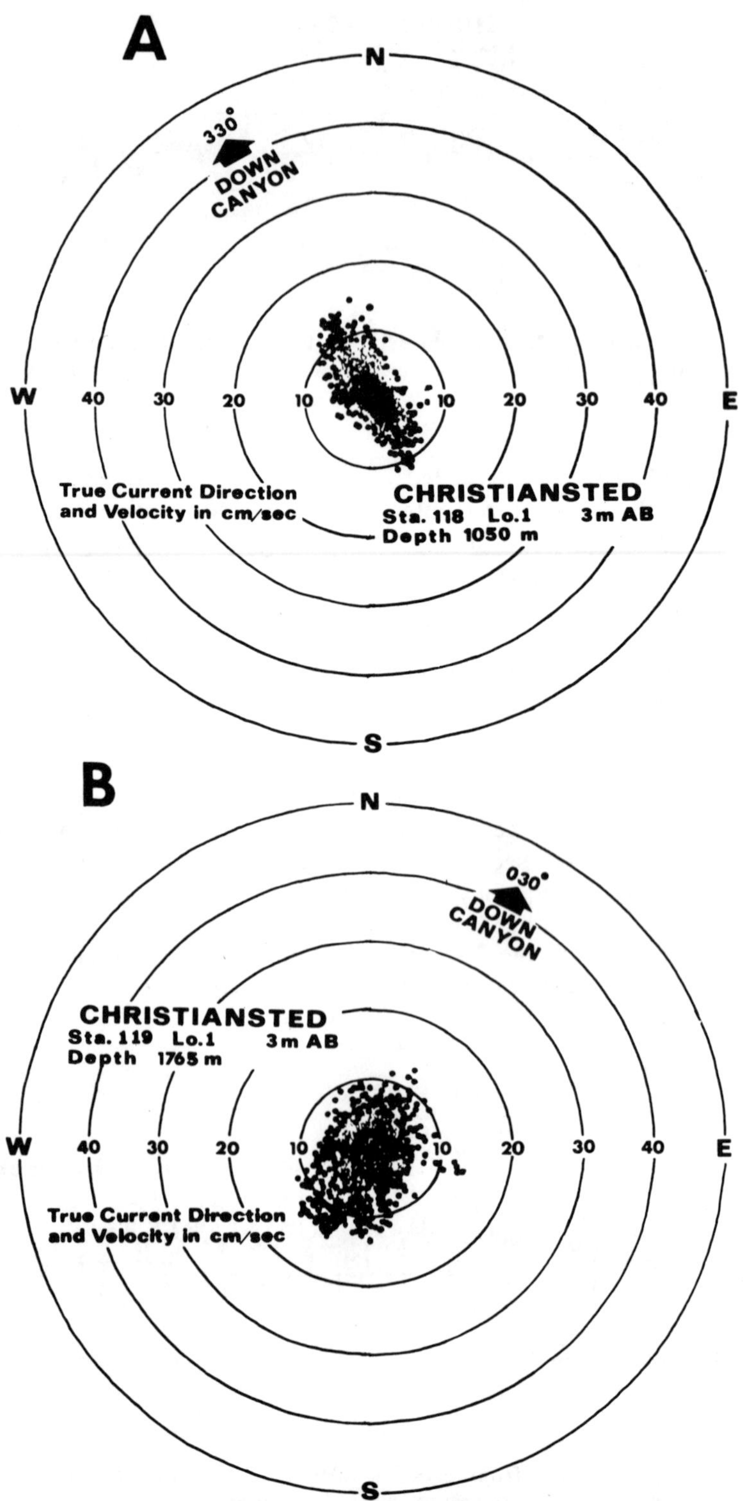
A
N
330°
DOWN CANYON
W
40
30
20
10
10
20
30
40
E
True Current Direction and Velocity in cm/sec
CHRISTIANSTED
Sta. 118 Lo.1 3m AB
Depth 1050 m
S
B
N
030°
DOWN CANYON
CHRISTIANSTED
Sta. 119 Lo.1 3m AB
Depth 1765 m
W
40
30
20
10
10
20
30
40
E
True Current Direction and Velocity in cm/sec
S

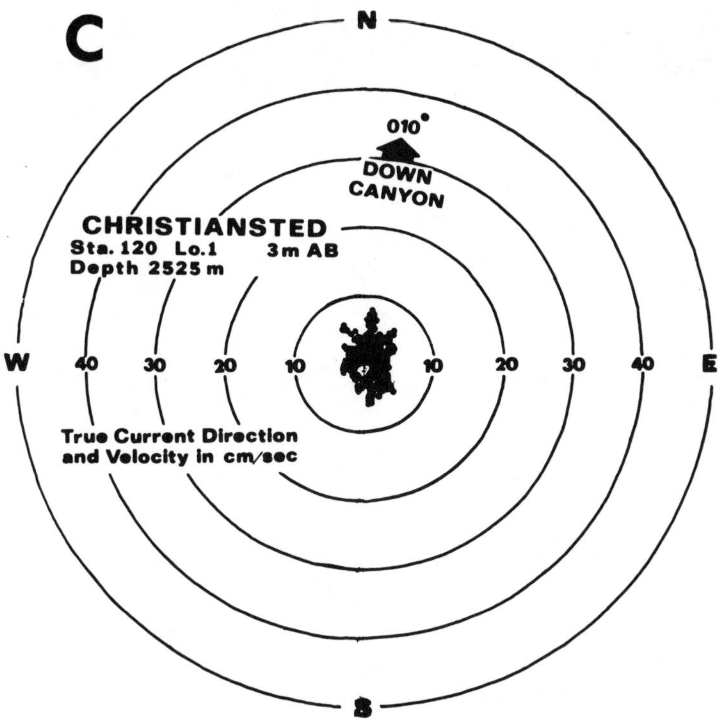

FIG. 108—Polar plots from records of the three outer stations in Christiansted Canyon. Because of the closely spaced data points in the inner circles, the patterns are shown as shaded areas. Notice slow speed of the currents in this canyon compared with the somewhat faster currents in other canyons where we have records at considerable depths.

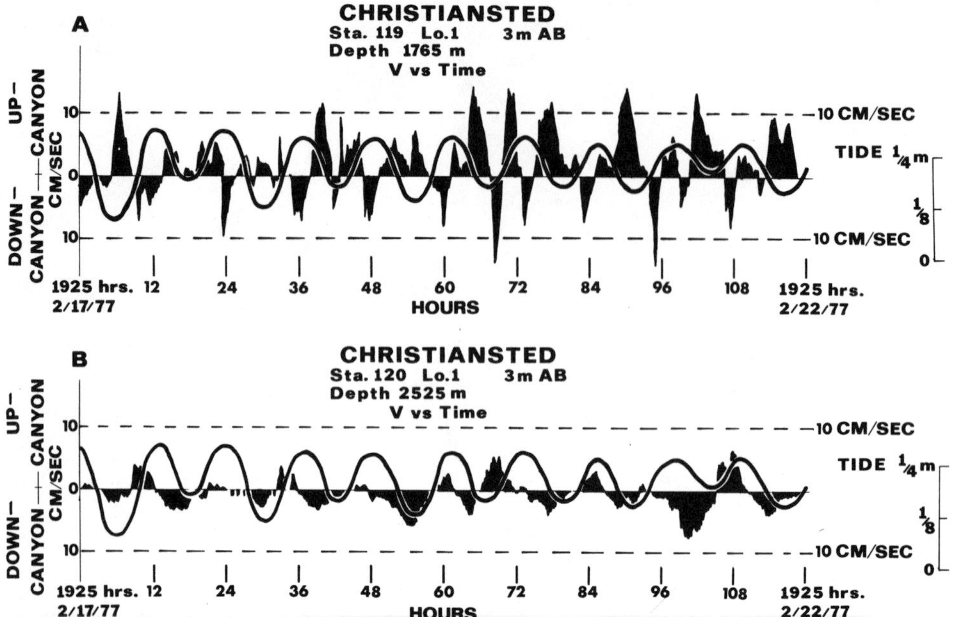

FIG. 109—Time-velocity curves at 1,765 and 2,525-m depths in Christiansted Canyon show that only the deepest station has current alternations that correspond roughly with the tides. Note weakness of the currents at the deepest station.

an increase in length with depth for these more normal canyon currents. Because of the extremely small tidal range, the effect of the semidiurnal tide is not indicated even at this considerable depth. The time-velocity plot (Fig. 109A) shows that the upcanyon flows are longer in duration than most of those in a downcanyon direction. An attempt to correlate events from this station with the time-velocity plot from that of the 1,050-m depth was unsuccessful. The velocities are about the same at this 1,765-m station with maxima of 14 cm/sec for upcanyon and 12.5 cm/sec for downcanyon. The net flow was upcanyon (Fig. 110), which is contrary to that at 3 m in the 1,050-m station. The polar plot indicates considerable crosscanyon current activity with some of the fastest currents being in that direction (Fig. 108B).

At a depth of 2,525 m the record for 3 m above bottom shows still slower currents (Fig. 109B) with maximum speed at 8 cm/sec downcanyon and 6 cm/sec upcanyon. Considerable crosscanyon flow is indicated, but at even slower speeds (Fig. 108C). The alternation of flow direction has at this considerable depth become close to tidal although agreement with the tide is not very perfect (Fig. 109B). The average length for a cycle is 12 hours. An attempt to match time-velocity curves with the 1,756-m depth record again failed to show any relationship. The progressive-vector plot shows net downcanyon flow, contrary to the 1,765-m station (Fig. 110B). Thus there is little consistency in net flow direction at the various depths in this canyon. At 30 m above bottom the polar plot is more directional than at 3 m, while the alternations of current flow direction are somewhat irregular but close to tidal in period. The average period is only 10 hours, largely because some of the flows slowed but did not completely reverse direction so that they continued in the same general direction while several others reversed only briefly.

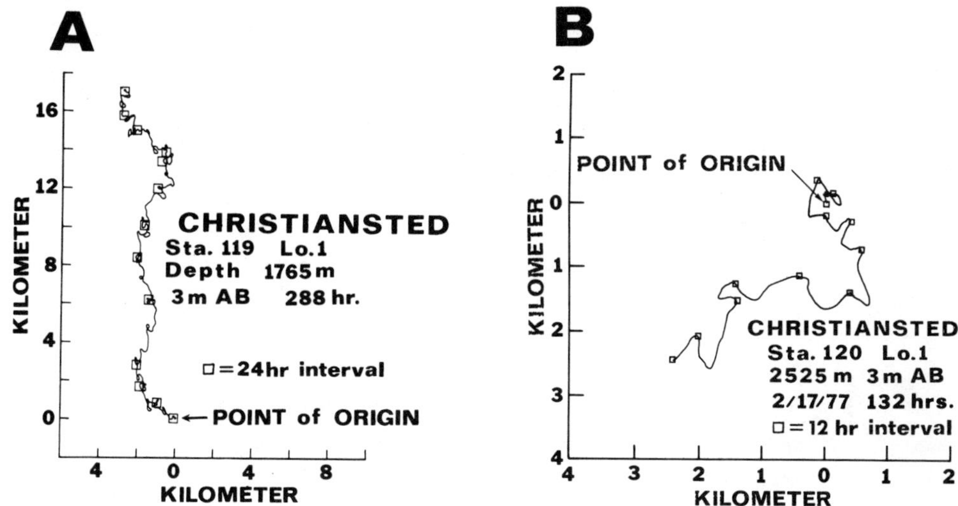

FIG. 110—Progressive-vector diagrams of the currents at the two deepest stations in Christiansted Canyons: (**A**) indicates upcanyon net flow at 1,765 m. (**B**) indicates downcanyon net flow with considerable crosscanyon flow at 2,525 m.

Thus the up- and downcanyon alternations of current flow become essentially tidal only at the 2,525-m depth, an even greater depth than at Rio Balsas. Since the tide range here is very small, the concept that great depth is required before tidal periods become involved corresponds here to what we have found elsewhere.

Pacific Island Canyons

We covered two areas among the Pacific Islands in our current-meter studies. Working with U.S. Navy Ships, we were able to operate in a mid-ocean environment off the north side of the Hawaiian Island of Kauai. Later we directed one leg of a Scripps Institution of Oceanography expedition in the Indo-Pacific where we had the opportunity to operate off northwestern Luzon, one of the island arc areas.

Kauai Canyons, Channels, and Seavalleys

In cooperation with research investigations of the U.S. Navy, we were given the opportunity to obtain some current-meter records near the Barking Sands Missile Tracking Range off northwest Kauai in the Hawaiian Islands. This work was arranged for us by P. J. Fischer and conducted as a cooperative program with him. We visited the area on two occasions on the U.S.S. *Silas Bent*, and later we obtained additional records when Gary Sullivan made several trips on a torpedo-recovery boat from the Barking Sands Base.

North of Kauai and Niihau Islands, the seafloor slopes seaward down to a depth of about 4,600 m (2,500 fm, Fig. 111). Directly off the Napali cliffs of Kauai, there are a series of submarine canyons, some of which extend seaward from land canyons. The submarine canyons terminate at a depth of about 2,200 m (Fig. 112). Off the channel between Kauai and Niihau, a broad slope valley (called Kaulakahi

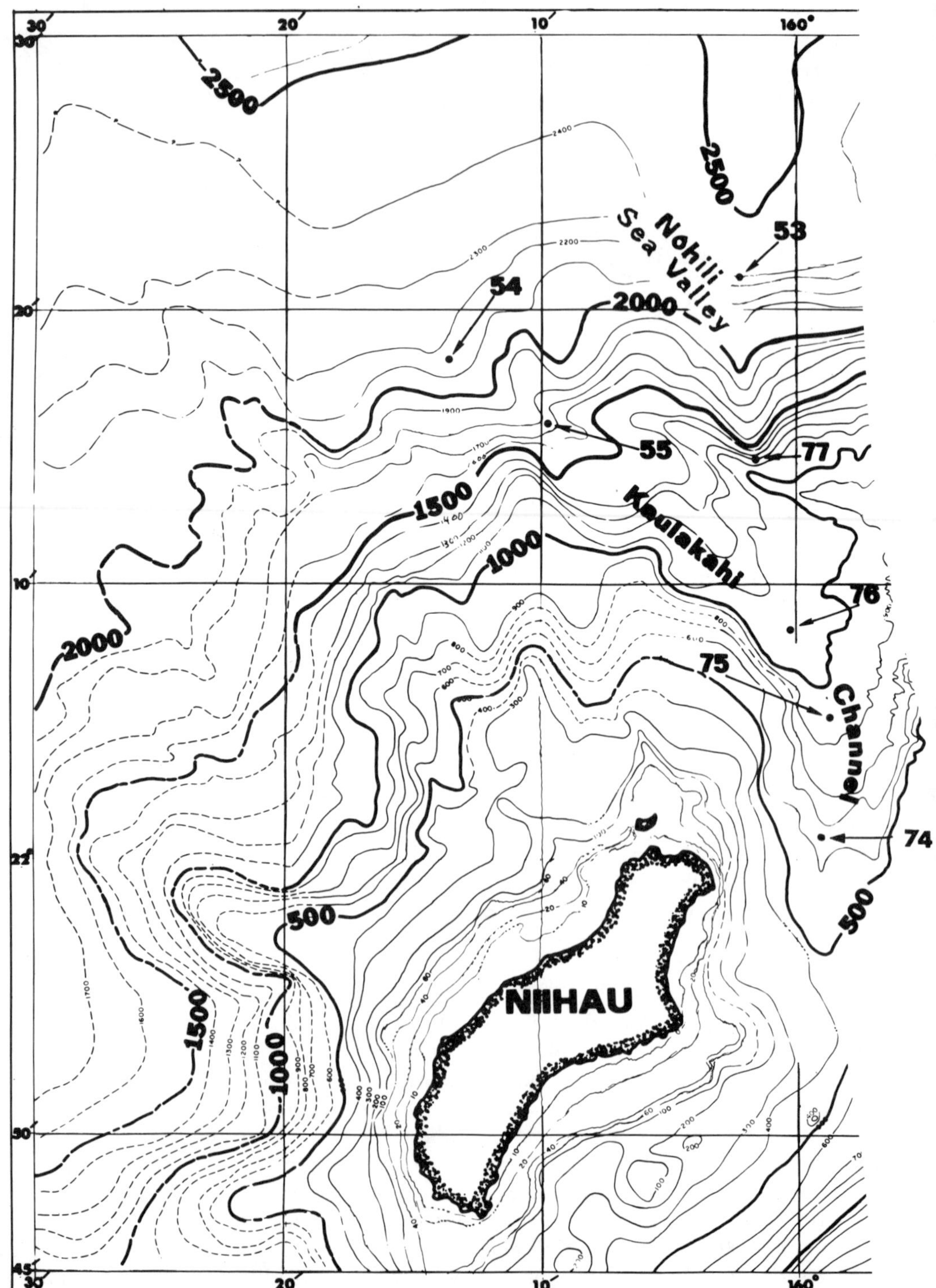

FIG. 111—Contour chart of the area north and west of Kauai showing the locations of

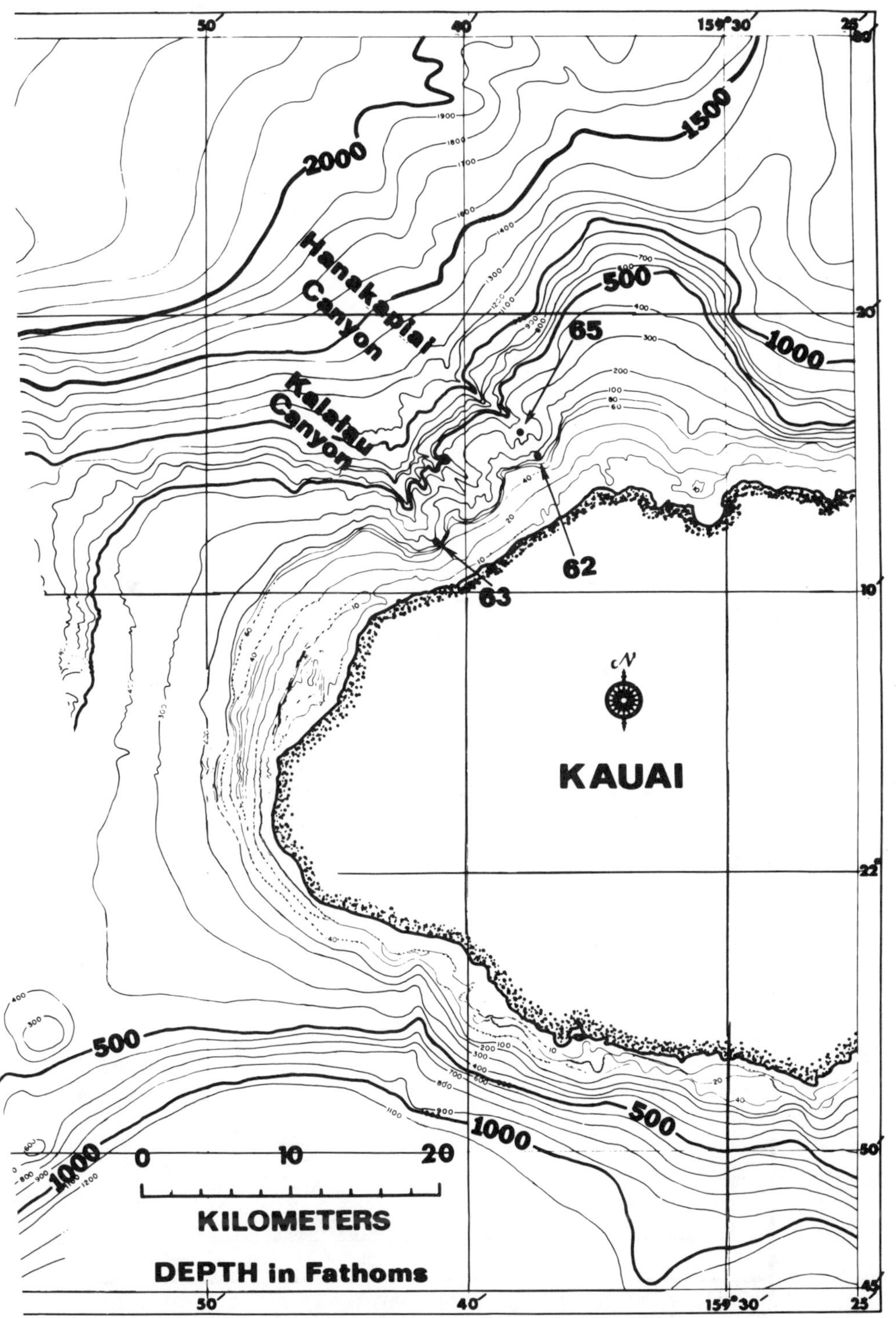

current-meter stations. From maps supplied by P. J. Fischer from U.S. Navy Surveys.

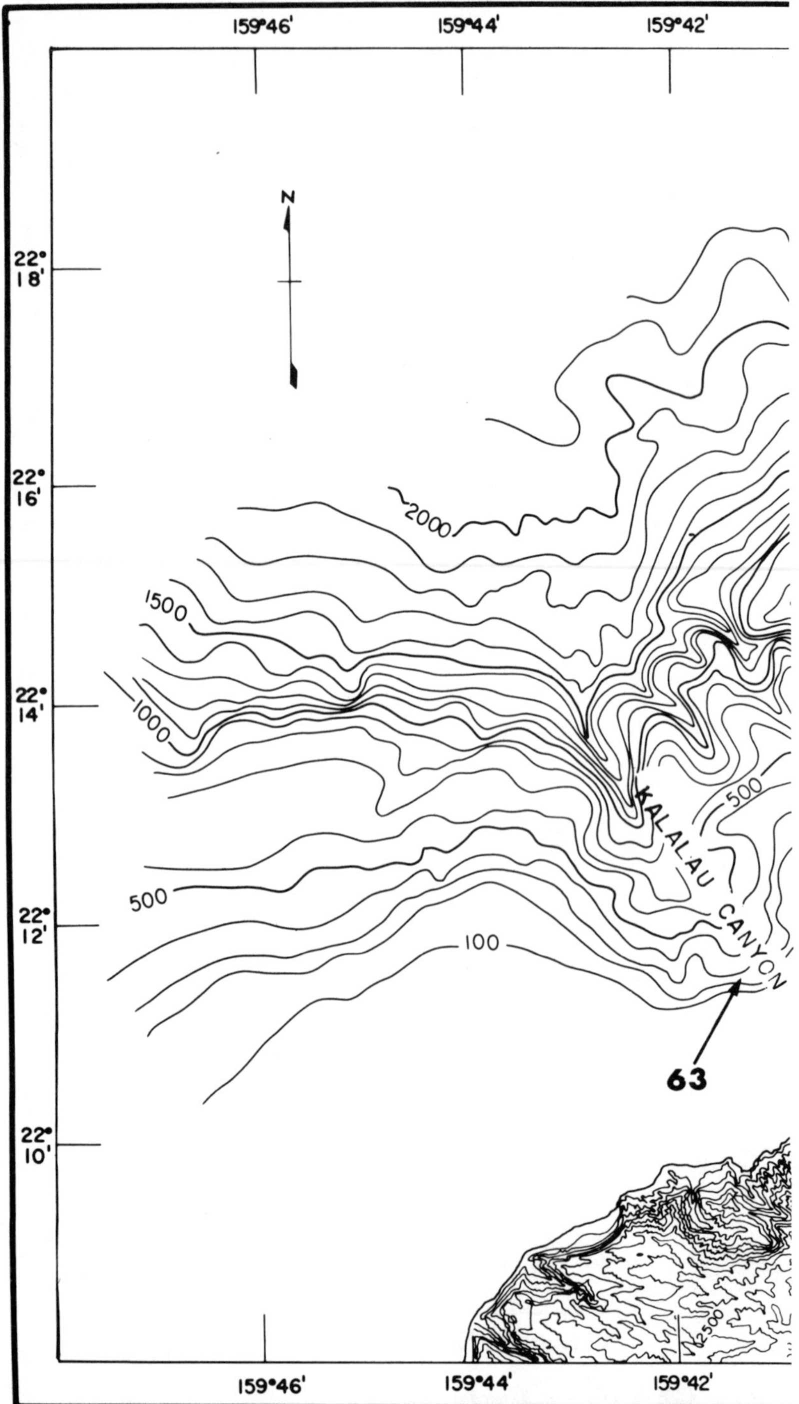

FIG. 112—Contours of the submarine canyons off the Napali Coast of Kauai showing relation to land canyons. Current-meter stations indicated. The fathograms taken at the time in the U.S. Navy Ship *Silas Bent* showed that Stations 62 and 63 were definitely at the bottom of canyons although this was

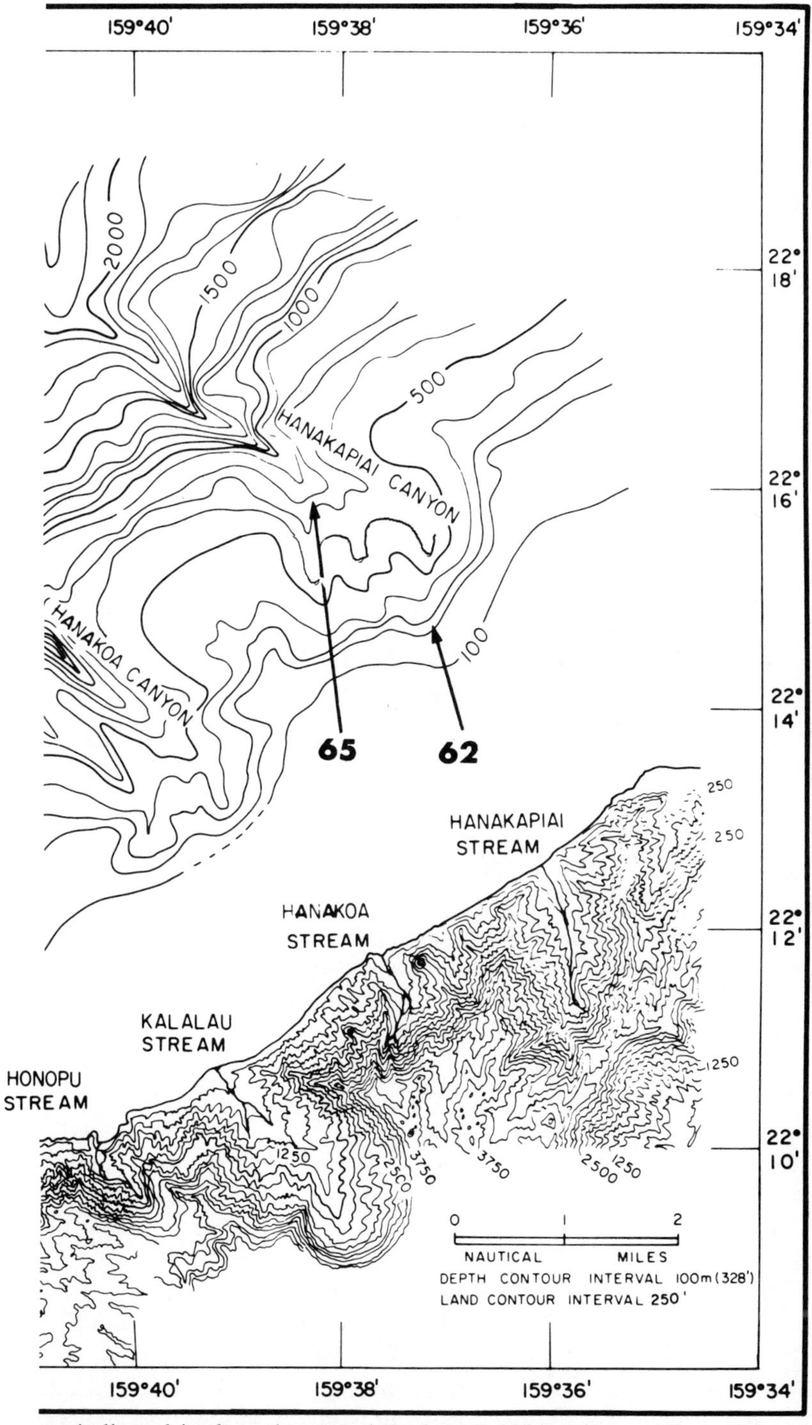

not indicated in the only map we had available based on a Shepard survey with inadequate equipment in 1962 (Shepard and Dill, 1966; Fig. 105). Navy survey has not been made available.

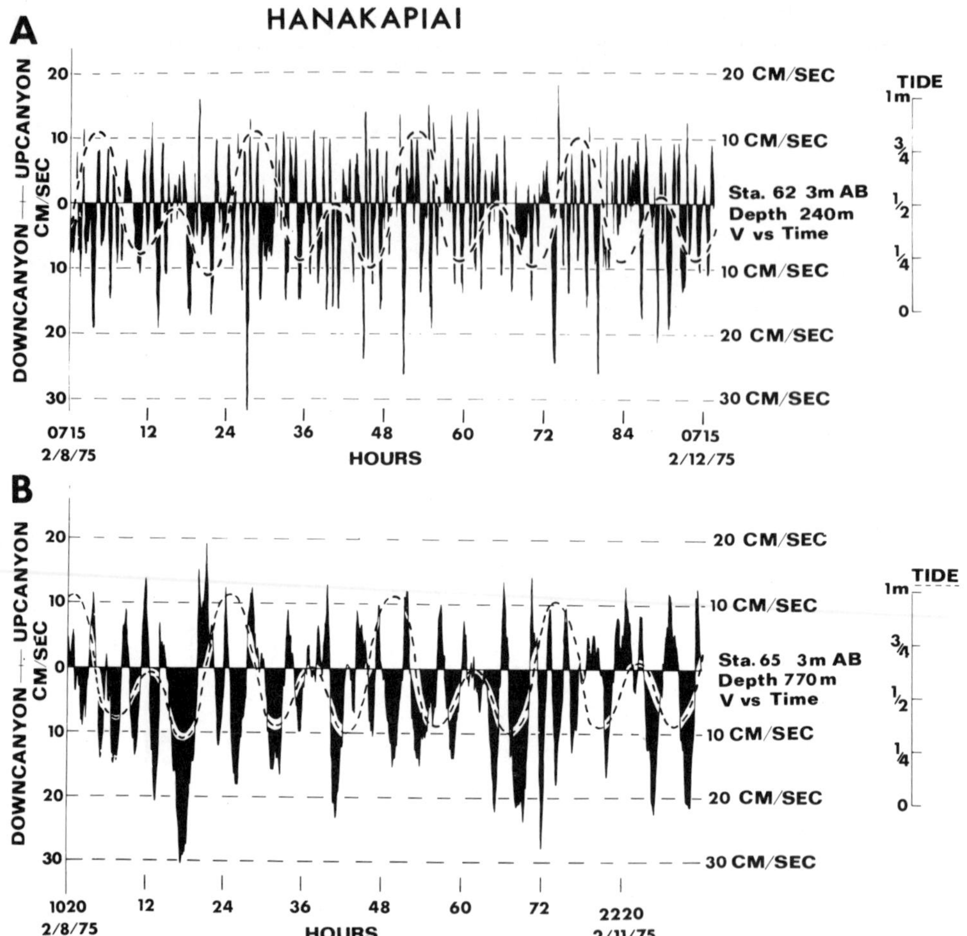

FIG. 113—Figure shows the irregular high-frequency alternations of current direction in the time-velocity curves at 240- and 770-m depths in Hanakapiai Canyon. Note increase in cycle periods at the deeper station. Tide is indicated by dashed line.

Channel) extends north, and then at a depth of 1,800 m (1,000 fm) bends to the northwest dropping to a depth of about 4,000 m (2,200 fm). A ridge extends west-northwest of Kauai, and on its north slope is Nohili Seavalley. It can be traced down to about 4,600 m (2,500 fm). These features were investigated, and the currents from the localities proved to be significantly different from others we have studied.

Canyons off the Napali Coast—The submarine canyons off Kauai appear to be related to the large land canyons that cut the cliff coast of northwest Kauai. These are among the largest land canyons in the Hawaiian Islands, and the adjacent submarine canyons also have considerable relief (Fig. 112). We obtained one current-meter record at a depth of 290 m in Kalalau Canyon, named for the largest of the north-coast land canyons, and two current-meter records in Hanakapiai Canyon at depths of 240 and 770 m. The currents in these submarine canyons show clearly the influence of the small tidal range, about 0.5 m, that characterizes the Hawaiian Islands. All of the time-velocity curves in the canyon show short flow periods between direction reversals (Fig. 113). At the 290-m depth in Kalalau Canyon the average cycle is about one hour. This alternation frequency compares favorably with the short oscillation cycles in San Lucas Canyon and at the heads of the St. Croix canyons where the tides are even smaller. At 240 m in Hanakapiai Canyon the average is slightly longer, about 1.3 hours. At 770 m depth, the cycle in Hanakapiai Canyon is distinctly longer, averaging 3.5 hours. still far different from the currents at similar depths in areas with moderate or large tide ranges.

The currents at the canyon heads are somewhat faster off Kauai than off the island of St. Croix, having maximum speeds at 240 m of 35 cm/sec downcanyon and 20 cm/sec upcanyon. At 290 m depth the currents are slower with a maximum of 17 cm/sec downcanyon and 12 cm/sec upcanyon, and at 770 m they reach 32 cm/sec downcanyon and 21 cm/sec upcanyon.

The polar plots are clearly related to the canyon axial trend with the best relation at 770 m (Fig. 114). All of the records show net downcanyon flow with fairly constant direction.

The Hawaiian tides have a strong diurnal component, one tide much larger than the other (Fig. 115). This does not appear to have any influence on velocities at the shallow stations in these canyons, but there is some tendency for speeds to increase at the 770-m depth during times of the larger tidal changes (Fig. 113B).

Channels and Deep Seavalleys—The deep, broad-floored valleys off the Kauai area have currents that are quite different from those in the canyons. The difference is due, in part, to their greater depths; although their broad floors, which contrast greatly with the narrow gorges of the canyons off the Napali cliffs, probably have a significant effect.

In Kaulakahi Channel we obtained records at depths of 1,188; 1,737; 1,939; 3,255; and 3,904 m. At 1,188 m we acquired direction only, as the speed did not record properly. However, direction changes seem to indicate that the alternation cycles are related to the tides at this depth. Most up- and downvalley cycles have a period close to 12 hours. At the 1,737-m depth the tidal relationships to the flows are evident (Fig. 115) with an average of 12.5 hours. These tidal-related direction changes are useful in indicating the advance of the internal waves along the valley axis. By comparing the direction record of the 1,188-m station and that of the 1,737-m station, we can find a fairly good match between the two. The time difference is approximately 5 hours and 45 minutes, with direction changes occurring earlier at the 1,737-m station with a distance of 8.1 km, suggesting upvalley advance of internal waves. Because we also had a third contemporaneous record station at 1,939 m, and here the records include both direction and speed, we can make a better match between the two deeper stations. Again the events occur earlier at the deep station (Fig. 116). The amount of offset is 5 hours in this case with a distance of 6.3 km, tending to confirm the less reliable matching of times of direction changes at the 1,188-m and 1,737-m stations. A comparison between two contemporaneous stations at 3,255 and 3,904 m depths in the outer part of the same valley also suggests upvalley advance for the internal waves with a time difference of 6 hours and 9 minutes with a distance of 8 km (Fig. 117). Thus the evidence for upvalley advance of internal waves is even better here than in La Jolla and Hueneme Canyons where three examples of upcanyon advance were recorded.

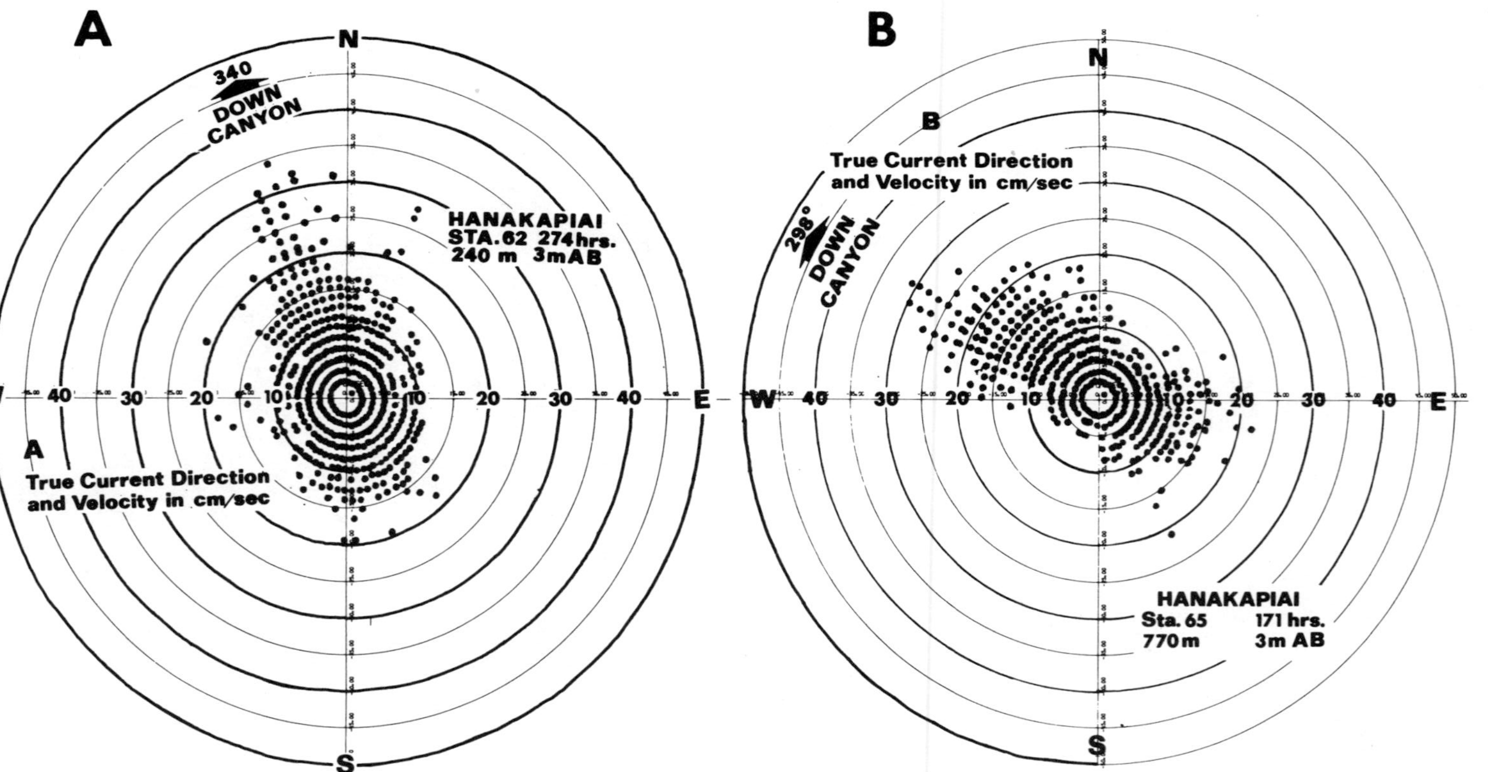

FIG. 114—Polar plots of records at 240 and 770-m depths in Hanakapiai Canyon off Kauai. Notice relation to canyon axis and greater strength of downcanyon currents.

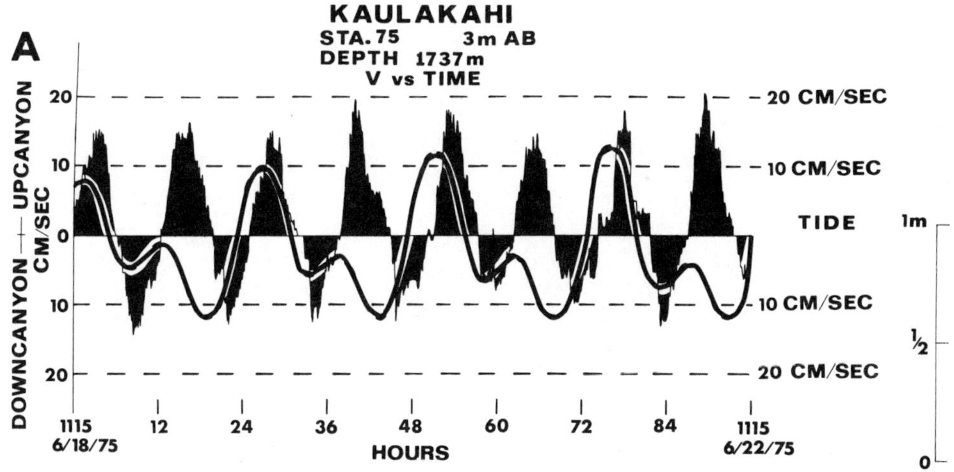

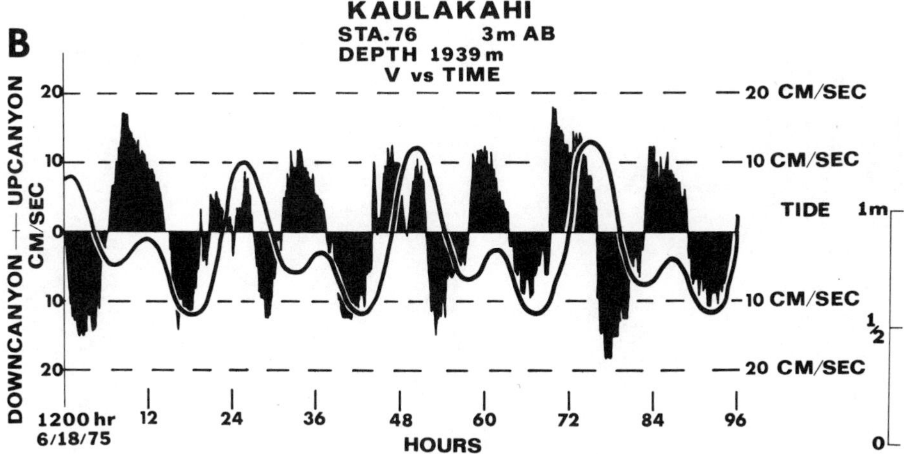

FIG. 115—**A**, time-velocity curve at 1,737-m depth in Kaulakahi Channel northwest of Kauai. Currents have semidiurnal tidal relation but strongest upcanyon flows appear to come directly after diurnal tides of small amplitudes. **B**, time-velocity curve at 1,939-m depth in Kaulakahi Channel with same relation to the small diurnal tides as at 1,737 m.

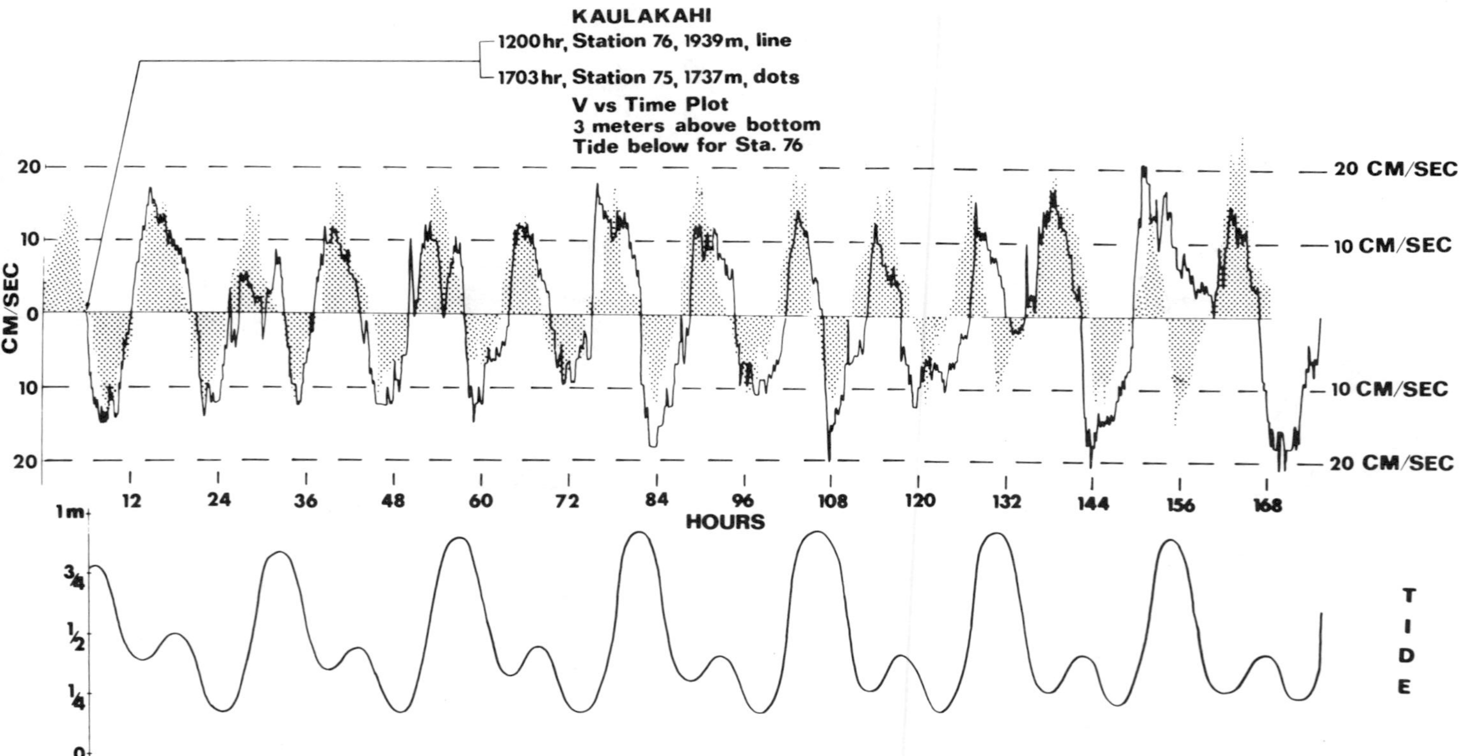

FIG. 116—Superposition of time-velocity curves at 1,737-m and 1,939-m depth in Kaulakahi Channel. Best fit puts time at the shallow station 5 hours later than that at deep, suggesting upvalley advance of the internal waves.

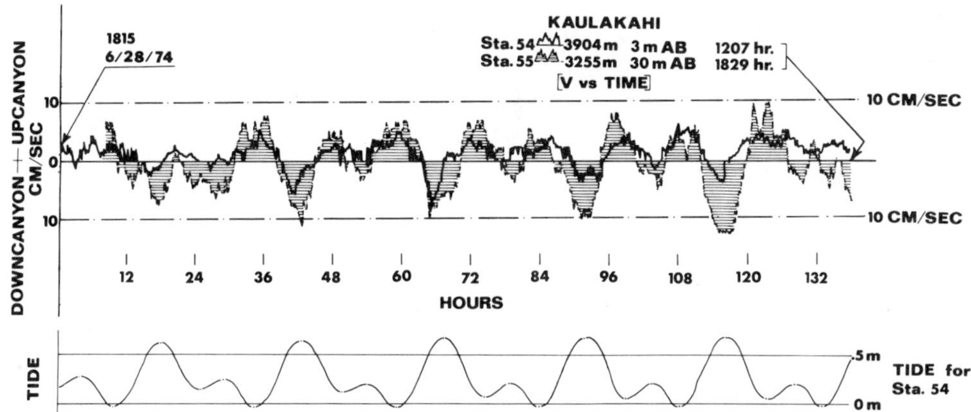

FIG. 117—Superposition of time-velocity curves at 3,255 and 3,904-m depths in Kaulakahi Channel. Best fit again shows the shallow station with later times than those at the deeper station (6 hours and 22 minutes), suggesting internal waves advancing upvalley here also. Only the record at 30 m above bottom is available at 3,904 m.

The velocities at the stations in the Kaulakahi Channel decrease with depth. The maximum current at 1,737 m is 27 cm/sec in two somewhat different directions, both roughly upvalley; at 1,939 m it is 21 cm/sec in both up- and downvalley directions; at 3,255 m with the velocities shown only at 30 m above bottom the maximum speed is 13 cm/sec downvalley; and for 3,904 m the maximum at 3 m above bottom is 9 cm/sec downvalley with some crosscanyon up to 10 cm/sec. Here again in Kaulakahi Channel we have very slow currents in deep water as in the deepest station in Christiansted Canyon.

At 1,737 m the currents, as indicated in the progressive-vector diagram, show a series of what are essentially tidal ellipses (Fig. 118A). This depicts the rotary tidal motion that occurs in many places on the continental shelf where flows are not topographically controlled (Shepard and Marshall, 1978). A somewhat similar series of ellipses is shown on the progressive-vector diagram for the 1,939-m station (Fig. 118B), but the sequence is less regular than at 1,737 m. The net flow at 1,737 m is upcanyon but at 1,939 m it is about even up- and downvalley. The polar plots for these stations in Kaulakahi Channel show very little relation to the axis of the valley, indicating considerable crossvalley flow (Fig. 119). They are somewhat similar to what was found at the deep station in Hudson Canyon (Fig. 88B) and in one of the Salt River Canyon records that was taken near the level of the canyon rim (Fig. 102B).

Records in Nohili Seavalley were obtained at the 2,213 m and 4,206 m depths; both 3 m above bottom. At the shoaler station the currents show a number of short cycles that are not tidal, but many of the direction changes are related to the tides (Fig. 120A). The average period of only 8 hours is not in agreement with the other deep-valley records from Kauai and elsewhere. The fastest currents are downvalley at about 13 cm/sec with net flow also downvalley.

At a depth of 4,206 m the valley is not well-developed, and we find extremely slow currents, as might be expected at this, the deepest of all our usable records. The

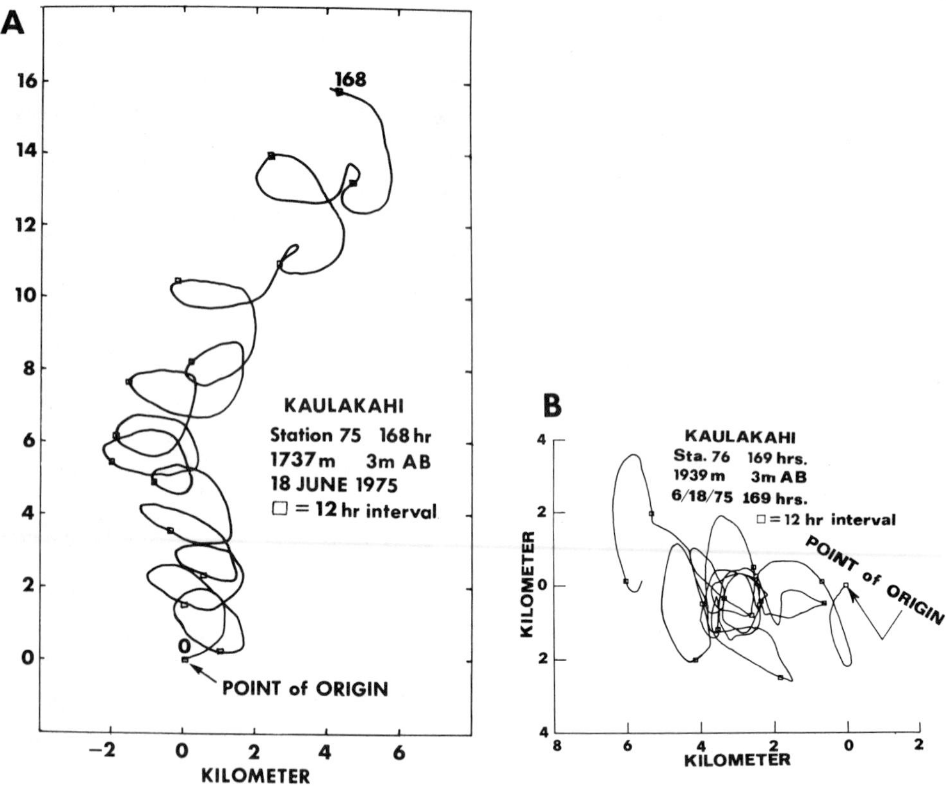

FIG. 118—Progressive vector diagram from Kaulakahi Channel at 1,737- and 1,939-m depths showing the elliptical path often characteristic of tidal currents on the continental shelf, suggesting that the channel is broad enough so that the valley walls do not appreciably confine the current motion or that the currents meander through the broad valley.

fastest speed is around 3 cm/sec in a downvalley direction, and the net flow is also downvalley. The time-velocity curve shows that the currents alternate in direction with an approximate relationship to the tides, but with some short-period, nontidal alternations (Fig. 120B). The average reversal period is about 10 hours. We had a current meter at 4,580 m, but the currents were too slow for an adequate record so we were unable to show the current graphically at this station.

Abra Delta, Northwest Luzon, Philippine Islands

In 1976, the Scripps Institution of Oceanography Expedition INDOPAC included an operation off the Abra delta of northwest Luzon. We had the opportunity to measure currents in the delta-front valleys as well as to make detailed bathymetric and seismic surveys of the area (Fig. 121).

The Abra delta is another locality where, as off Rio Balsas, a relatively large delta has built across a continental shelf so that it is now building the slope forward into a relatively deep basin. The delta-front slope is creased by many seavalleys in contrast to the slope directly north and south of the delta where seavalleys are less common. Directly seaward of the delta front is a fault trough. West of this valley a

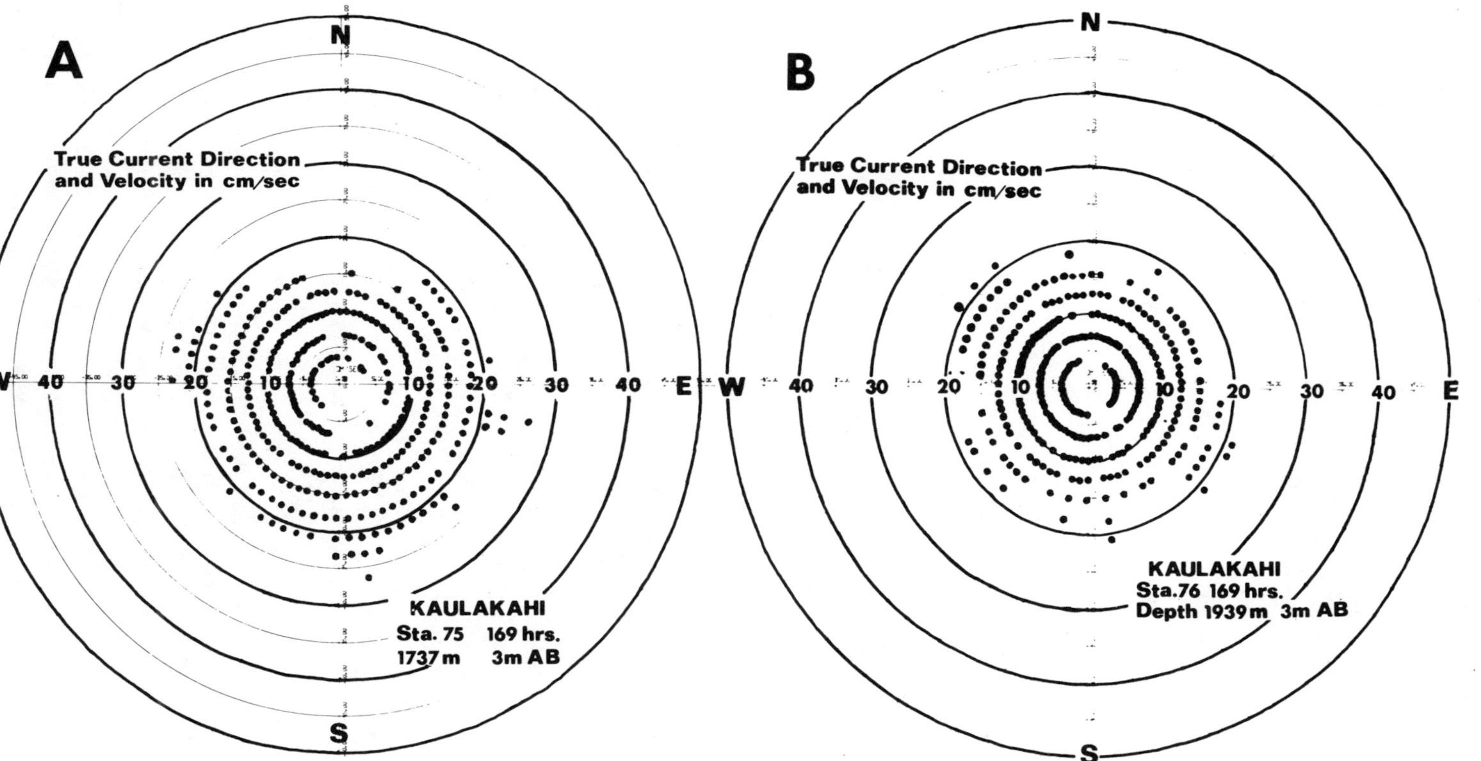

FIG. 119—Polar plots of currents at 1,737- and 1,939-m depths in Kaulakahi Channel showing again that there is little if any confinement of current to the axis of the valley in either case.

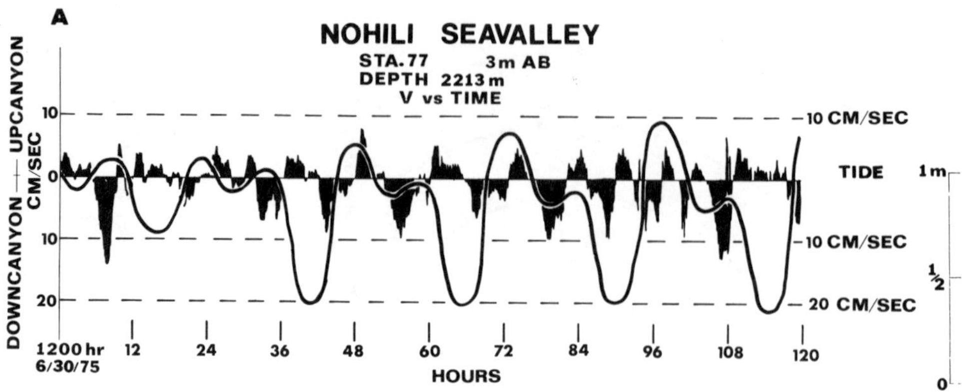

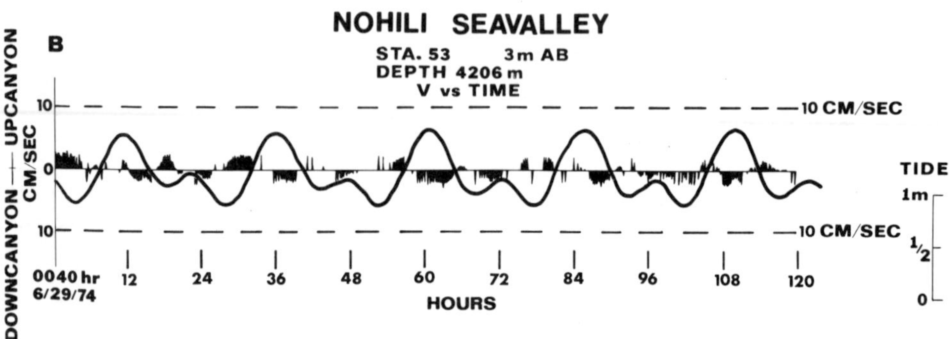

FIG. 120—Time-velocity curves at 2,213- and 4,206-m depths in Nohili Seavalley off northwest Kauai. Currents are usually faster at the 2,213-m station when the diurnal tide changes were small; this was also true in Kaulakahi Channel. Note the weak currents at 4,206 m which is the deepest station of all those where we obtained records.

shallow bank separates the area from the slope leading to the Manila Trench. The abrupt landward slope of the shallow bank is evidently a fault scarp (Fig. 121). Seismic profiles show the reality of the fault. It is very likely a continuation of the great Philippine fault which is thought to leave the coast just to the south.

Abra Seavalley—Current-meter records were obtained directly off the Abra River mouth in Abra Seavalley. It has walls 200 to 300 m high and a winding course. Natural levees are indicated on both sides. Our records had current directions in all cases, but only two of the current meters recorded velocities. We obtained good records from the axis at a 622-m depth at both 3 and 30 m above the bottom (Fig. 24). The tide had a range of about 0.9 m, somewhat intermediate in comparison to the other areas where we have records. Although the tide was essentially diurnal at the time, the time-velocity plot (Fig. 24) shows that the peak velocities both up- and downvalley are of a semidiurnal period. The plots show an average length of the up- and downvalley cycles of 11.5 hours, mostly suggesting semidiurnal tide. This contrasts with a 3.5-hour average at the 210-m station. The peak velocities at the 622-m station occurred diurnally. The downvalley flows are distinctly more irregular than those flowing upvalley. It seems clear that the one very fast flow is not related to the tide, but this will be discussed later. The net flow in the progressive-vector diagram is downvalley with a rather constant direction. The polar plot shows that most of the flows are along the axis despite some divergence northward in the case of the turbidity current (Fig. 122).

The record at 30 m above bottom at the 622-m station in Abra Seavalley has features similar to those at 3 m (Fig. 24B). The flows show about the same relation to the tides, but curiously the fast upcanyon flows are related to diurnal tides and downcanyon flows to semidiurnal, or are at least more frequent. Net flow in the progressive-vector diagram also is downvalley with some divergence to the north of the axis for the fastest flows. Velocities are slightly less than at 3 m. The same close relation of flows to the axis is indicated as at 3 m above bottom (Fig. 122B).

Turbidity Current—There seems to be little doubt but what both the 3-m and 30-m records at the 622-m depth station in Abra Seavalley show the existence of a turbidity current (Fig. 24). The fast downvalley flow that occurred simultaneously on both instruments at about 79 hours after the records started is in decided contrast to all previous currents in these records, reaching as it does 73 cm/sec (using one-minute measurement intervals) at 3 m above bottom and 53 cm/sec at 30 m. In both records the downvalley flow was preceded by a fast upvalley flow as in our other turbidity-current records. Following the surge there was a long period during which no current was observed. Both current meters were coated with fine sand when retrieved, indicating significant transport, and that the turbidity current had a thickness in the water column of at least 30 m. Unfortunately our meter at the shoaler station did not record speed.

Pinget Seavalley—Pinget Seavalley is located off Pinget Island at the north end of the Abra delta. This seavalley has well-developed levees along most of its length and has little if any indication of erosion. Walls are mostly less than 100 m high. In Pinget Seavalley we obtained rather unsatisfactory records at depths of 321 m and 1,350 m. At 321 m the 3-m record showed only very slow currents. The direction alternates very irregularly but with an average period of about one hour showing the effect of the small tide range. It is quite possible the occasional speed does not give the true picture, as there are too many gaps in this record which suggest malfunctioning of the instrument. At 1,350 m, only the 30-m above bottom record had both speed and direction. The time-velocity curve at 30 m (Fig. 9) shows cycles of direction alternation that are close to those of the semidiurnal tides, although the tide table only indicates diurnal tides at the time of the record. The major peaks of upcanyon flow are separated by over 12 hours, although the average cycle is only 10 hours. The fastest flows are downvalley and the net flow also is in that direction.

Congo Canyon

The submarine canyon at the mouth of the Congo River is somewhat unique in that it penetrates 28 km into the Congo estuary. The length and depth of this estuary is particularly surprising because the Congo is the largest river in Africa, and it introduces an enormous amount of sediment at its mouth. The significance of the large sediment discharge is indicated by the fact that it has proven impossible to maintain a submarine cable along the coast across this submarine canyon outside the estuary (Heezen et al, 1964). After each of the three attempts the cable was carried away, apparently by turbidity currents during seasons of large river discharge. As further indication of the sediment load, we found an enormous submarine fan beyond the submarine canyon with several distributary fan valleys extend-

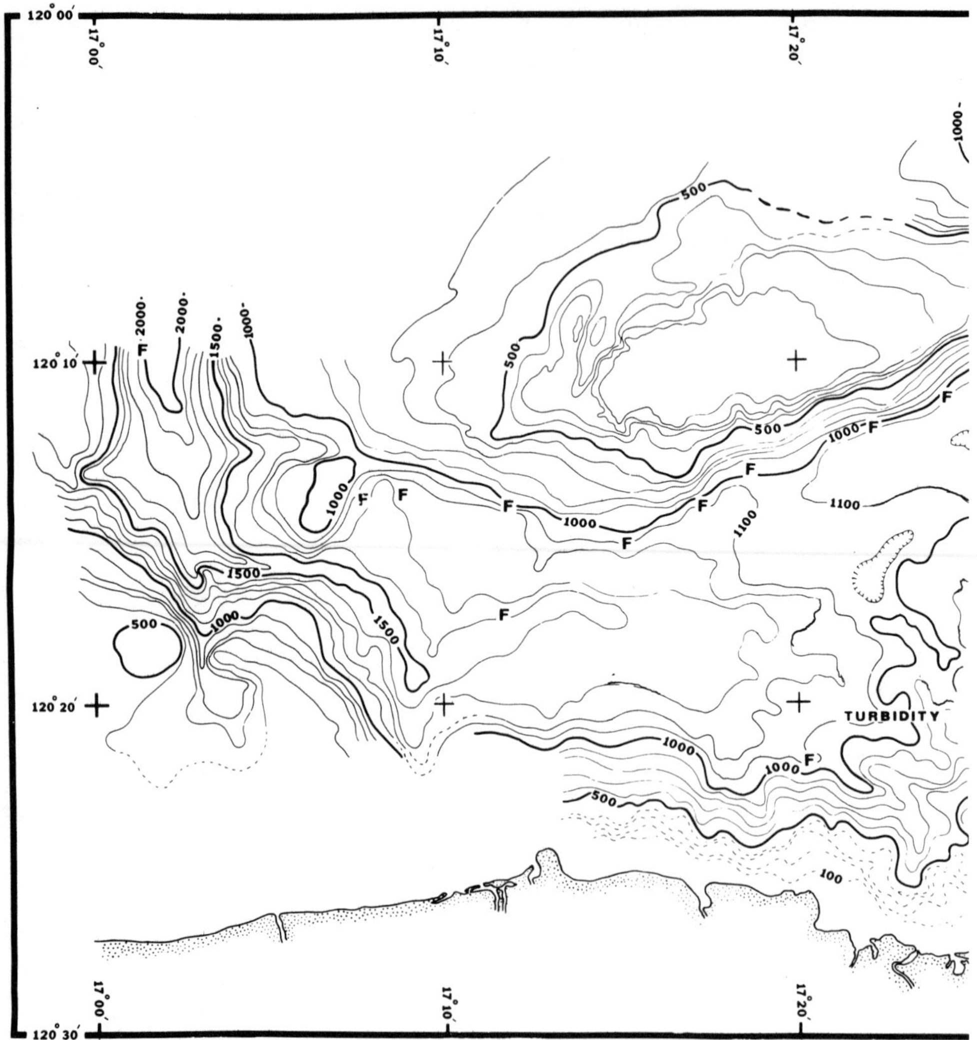

FIG. 121—The seafloor topography off the Abra delta, showing the location of the current-meter stations. Notice numerous seavalleys off the delta. Turbidity current occurred at station 104. From a survey on Scripps Institution *Thomas Washington*; contours by the writers.

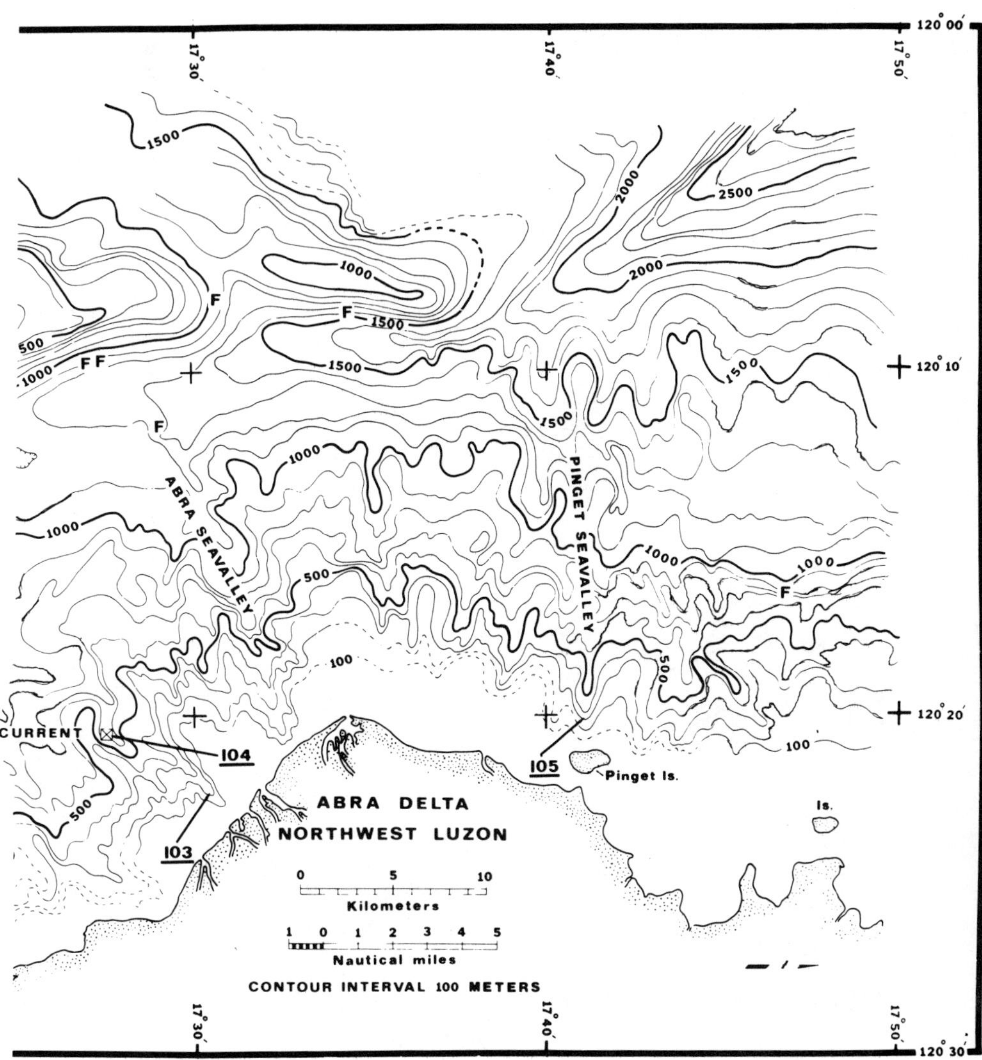

Seismic profiles indicated faults underlying the bottom at positions indicated by **F**. Note relation to landward margin of bank.

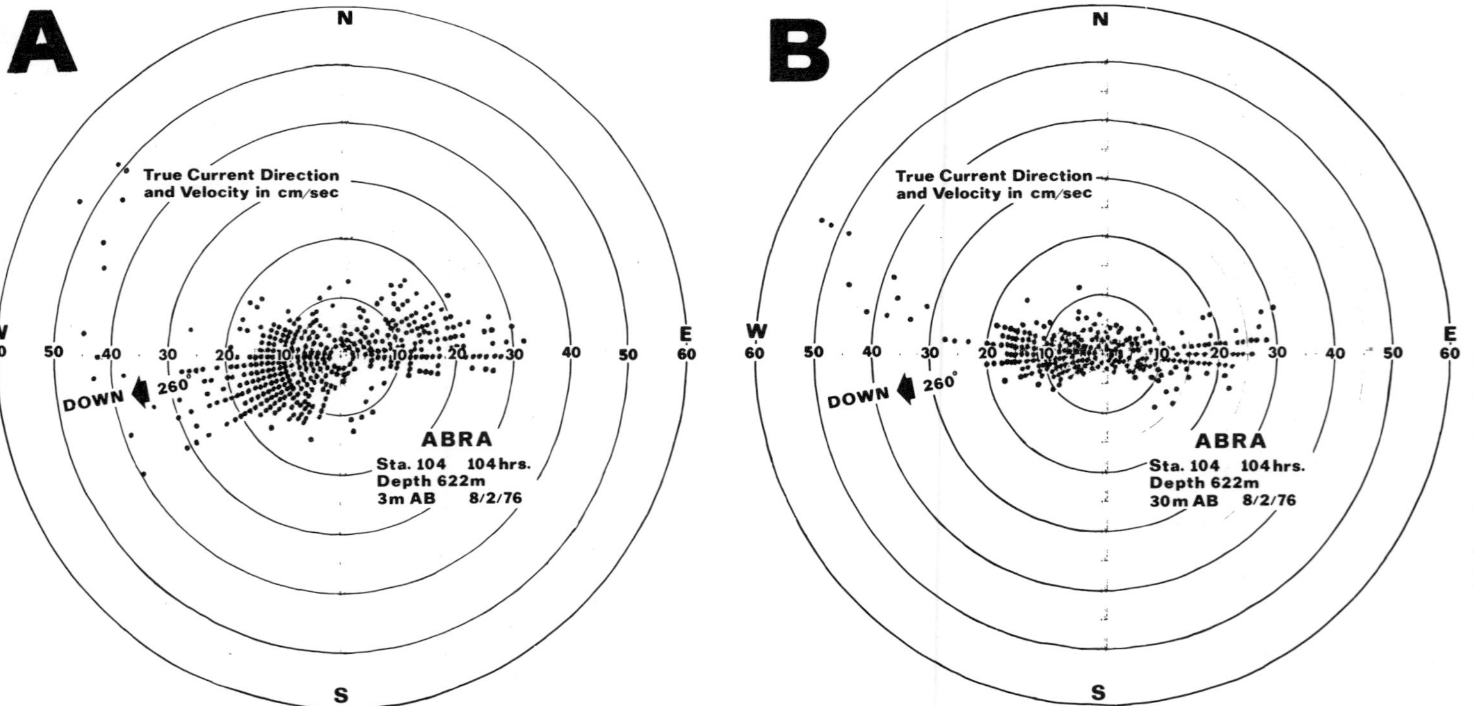

FIG. 122—Polar plot of the currents at 622-m depth in Abra Canyon showing good relation to the canyon axis of most currents, but divergence of about 30° to the north for the turbidity current at both 3 and 30 m above bottom.

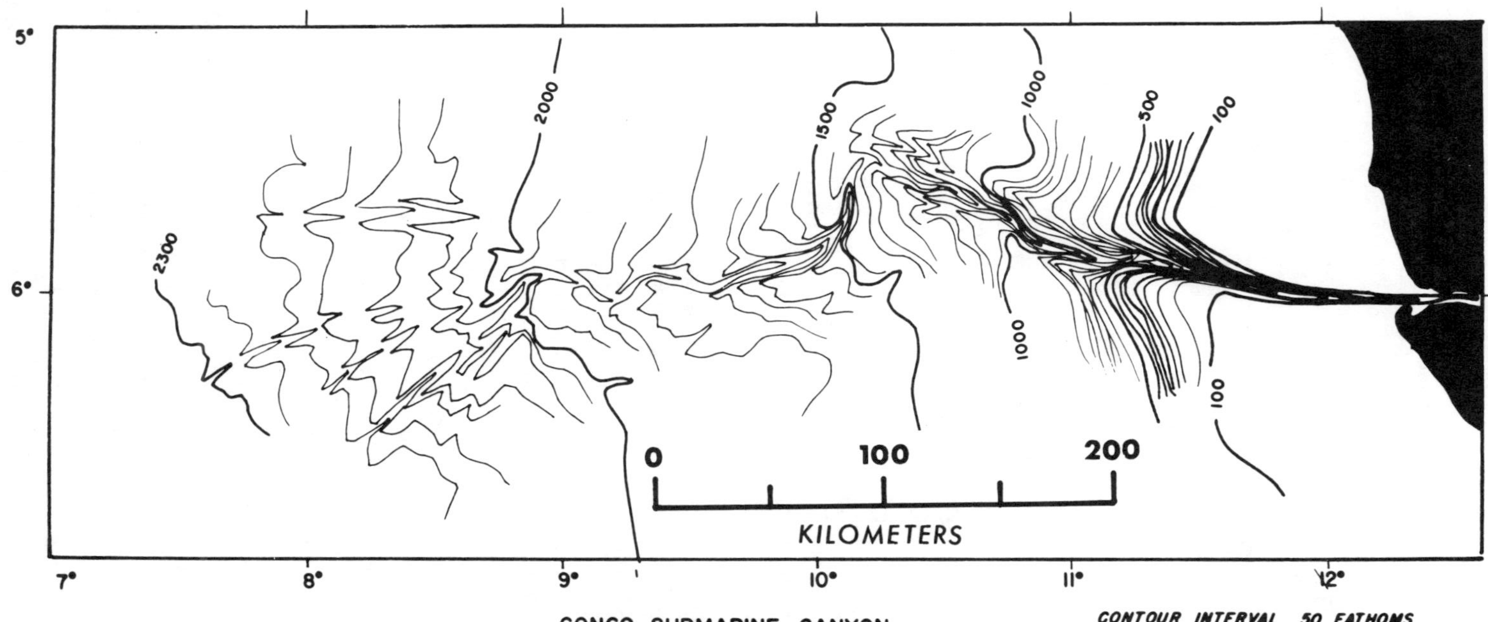

FIG. 123—Contour chart of Congo Canyon showing the valley distributaries and the inner part of the fan that is seaward of Congo Canyon (Shepard and Emery, 1973).

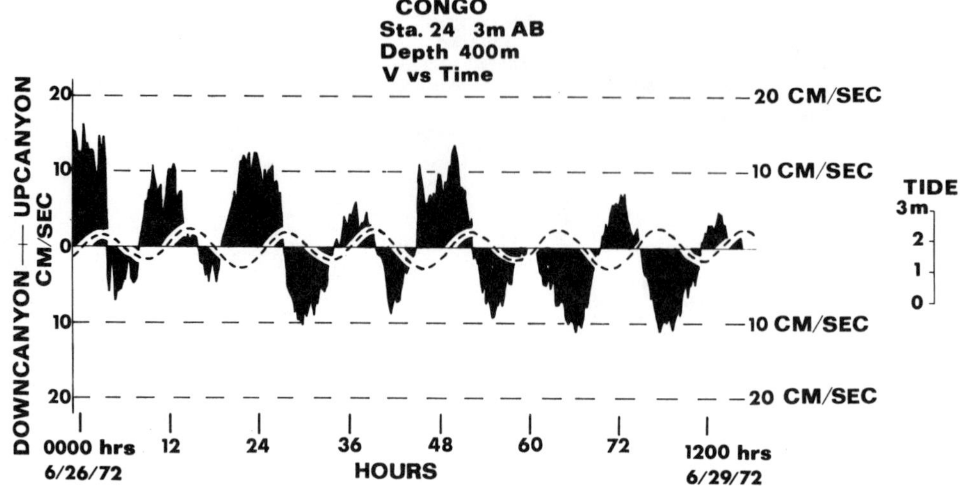

FIG. 124—Time-velocity curve of record at 400-m depth in Congo Canyon. One of the two rare cases where current flow agrees with phases of tide. Tide is indicated by dashed line.

ing beyond the canyon 750 km seaward across this great fan (Fig. 123).

In 1972 we had an opportunity to obtain a current-meter record in Congo Canyon during an expedition on the *Atlantis II* of Woods Hole Oceanographic Institution with K. O. Emery as chief scientist. Our record was from 400-m axial depth where the canyon now appears to have a fairly flat floor (Shepard and Emery, 1973).

The time-velocity curve from the Congo is compared with a tide record supplied us by the Gulf Oil Corporation, which has a station on an island just south of where we dropped our current meter. An interesting relationship is found between the tide and the time-velocity curve. The currents are moving upcanyon almost all of the time that the tide is rising and mostly flowing downcanyon while the tide is ebbing (Fig. 124). Our only other record where such a relationship is found is from the 85-m station off the Fraser delta where an unusually large river is emptying. The Fraser River is in marked contrast to the Congo, because a delta has been built and there is no deep estuary such as exists at the Congo River mouth.

The alternations of up- and downcanyon flow are clearly related to the semidiurnal tide. The range of the tide is 1.2 m, which is about intermediate for the areas in which we have worked. It is therefore rather surprising that such a relationship between the tides and current-flow direction alternations exists at this rather shallow depth.

The progressive-vector diagram (Fig. 125 A) shows that the net flow was upcanyon, perhaps not surprising since we could expect an intrusive salt wedge to move under the seaward flow of the low-salinity surface water off this large river. The strongest bottom flow, 22 cm/sec, was to the southeast during the first part of the record, which is directly opposite to the northwest flow of the surface current that flows along this part of Africa's coast. The polar plot (Fig. 125 B) shows that the bottom flows were generally northwest and southeast whereas the axis is believed to run mostly east-west in this part of the canyon, although we do not have

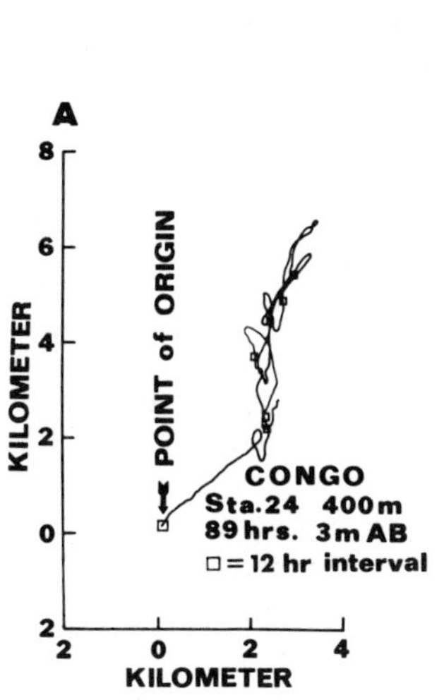

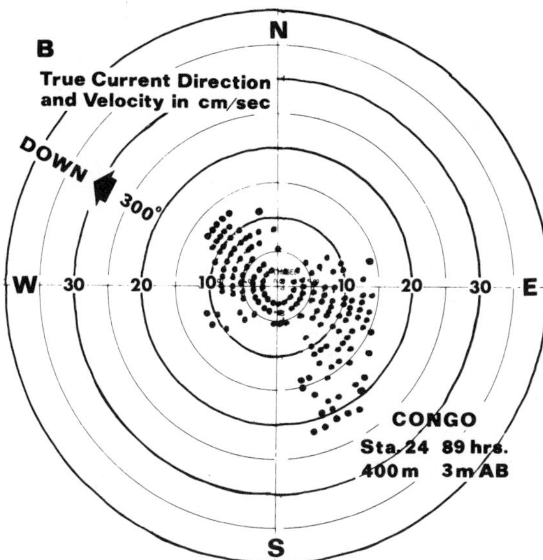

FIG. 125—**A**, progressive-vector diagram at Congo Canyon at 400-m depth shows diagonal crosscanyon flow at start. **B**, polar plot of record at 400-m depth im Congo Canyon. According to currents, axis is north of the east-west trend indicated by the chart.

sufficient soundings to show that this is the case. The old surveys are probably no longer accurate in view of the great amount of deposition and erosion that apparently takes place in this canyon.

REFERENCES CITED

Beer, R. M., and D. S. Gorsline, 1971, Distribution, composition, and transport of suspended sediment in Redondo Submarine Canyon and vicinity (California): Marine Geology, v. 10, p. 153-175.

Cacchione, D. A., G. T. Rowe, and A. Malahoff, 1978, Submersible investigation of outer Hudson Submarine Canyon, chapter 4, *in* D. J. Stanley and G. Kelling, eds., Sedimentation in submarine canyons, fans, and trenches: Stroudsburg, Pa., Dowden, Hutchinson & Ross, Inc., p. 42-50.

Coleman, J. M., and L. E. Garrison, 1977, Geological aspects of marine slope stability, northwestern Gulf of Mexico: Marine Geotechnology, v. 2, p. 9-41.

Daly, R. A., 1936, Origin of submarine "canyons": Am. Jour. Science, series 5, v. 31, p. 401-420.

Dill, R. F., 1964, Contemporary submarine erosion in Scripps submarine canyon: Ph.D. thesis, Univ. California, San Diego, 269 p.

Drake, D. E., and D. S. Gorsline, 1973, Distribution and transport of suspended particulate matter in Hueneme, Redondo, Newport, and La Jolla submarine canyons, California: Geol. Soc. America Bull., v. 84, p. 3949-3968.

Emery, K. O., and J. Hülsemann, 1963, Submarine canyons of Southern California, part I. Topography, water, and sediments: Los Angeles, Univ. Southern California Press, 80 p.

——— and E. Uchupi, 1972, Western North Atlantic Ocean: topography, rocks, structure, water, life, and sediments: AAPG Memoir 17, 532 p.

Felix, D. W., and D. S. Gorsline, 1971, Newport submarine canyon, California: an example of the effects of shifting loci of sand supply upon canyon position: Marine Geol., v. 10, p. 177-198.

Gennesseaux, M., P. Guibout, and H. Lacombe, 1971, Enregistrement de courants de turbidité dans la vallée sous-marine du Var (Alps-Maritimes): Acad. Sci. Comptes Rendus, Series D, v. 273, p. 2456-2459.

Gordon, R. L., and N. F. Marshall, 1976, Submarine canyons: internal wave traps?: Geophys. Research Letters, v. 3, p. 622-624.

Gorsline, D. S., and D. J. P. Swift, eds., 1977, Continental shelf sediment dynamics: a national overview: report of a workshop, Vail, Colorado, November 2-6, 1976 (sponsored by IDOE, NOAA, USGS, and ERDA), 134 p.

Greene, H. G., 1970, Geology of southern Monterey Bay and its relationship to the ground water basin and salt water intrusion: U.S. Geological Survey Open File Report.

——— 1977, Geology of the Monterey Bay region: U.S. Geological Survey Open-File Report 77-718, 347 p.

Heezen, B. C., and M. Ewing, 1952, Turbidity currents and submarine slumps, and the 1929 Grand Banks earthquake: Am. Jour. Science, v. 250, p. 849-873.

——— et al, 1964, Congo submarine canyon: AAPG Bull., v. 48, p. 1126-1149.

Hjulstrøm, F., 1939, Transportation of detritus by moving water, *in* P. D. Trask, ed., Recent marine sediments: SEPM Spec. Pub. 4, p. 5-31.

Inman, D. L., 1963, Sediments, chapter 5, *in* F. P. Shepard, Submarine geology (2nd ed.): New York, Harper & Row, p. 138-140.

——— 1970, Strong currents in submarine canyons (abs.): Trans. Am. Geophysical Union, v. 51, p. 319.

——— C. D. Nordstrom, and R. E. Flick, 1976, Currents in submarine canyons: An air-sea-land interaction: Ann. Review of Fluid Mechanics, v. 8, p. 275-310.

Isaacs, J. D., et al, 1966, Near-bottom currents measured in 4 kilometers depth off Baja California coast: Jour. Geophys. Research, v. 71, p. 4297-4303.

Jenkins, O. P., 1973, Pleistocene Lake San Benito: California Geology, v. 26, p. 151-163.

Keller, G. H., and F. P. Shepard, 1978, Currents and sedimentary processes in submarine canyons off the northeast United States, chapter 2, *in* D. J. Stanley and G. Kelling, eds., Sedimentation in submarine canyons, fans, and trenches: Stroudsburg, Pa., Dowden, Hutchinson & Ross, Inc., p. 15-32.

Kelling, G., H. Sheng, and D. J. Stanley, 1975, Mineralogical composition of sand-size sediment on the outer margin off the Mid-Atlantic states: Assessment of the influence of the ancestral Hudson and other fluvial systems: Geol. Soc. America Bull., v. 86, p. 853-862.

Kuenen, P. H., 1937, Experiments in connection with Daly's hypothesis on the formation of submarine canyons: Leidsche Geol. Med., v. 8, p. 327-351.

LaFond, E. C., 1962, Internal waves, *in* M. N. Hill, ed., The sea: New York, Interscience, v. 1, p. 731-763.

Lambert, A. M., K. R. Kelts, and N. F. Marshall, 1976, Measurements of density underflows from Walensee, Switzerland: Sedimentology, v. 23, p. 87-105.

Marshall, N. F., 1975, The measurement and analysis of water motion in submarine canyons, *in* E. Herg, ed., Ocean 75: New York, IEEE, p. 351-356.

——— 1978, A large storm-induced sediment slump reopens an unknown Scripps submarine canyon tributary, chapter 7, *in* D. J. Stanley and G. Kelling, eds., Sedimentation in submarine canyons, fans, and trenches: Stroudsburg, Pa., Dowden, Hutchinson & Ross, Inc., p. 73-84.

Martin, B. C., and K. O. Emery, 1967, Geology of Monterey Canyon, California: AAPG Bull., v. 51, p. 2281-2304.

Mathews, W. H., and F. P. Shepard, 1962, Sedimentation of Fraser River delta, British Columbia: AAPG Bull., v. 46, p. 1416-1438.

Miller, M. C., I. N. McCave, and P. D. Komar, 1977, Threshold of sediment motion under unidirectional currents: Sedimentology, v. 24, p. 507-527.

Moore, D. G., 1969, Reflection profiling studies of the California continental borderland: structure and quaternary turbidite basins: Geol. Soc. America Spec. Paper 107, 142 p.

Normark, W. R., and F. H. Dickson, 1976a, Sublacustrine fan morphology in Lake Superior: AAPG Bull., v. 60, p. 1021-1036.

——— ——— 1976b, Man-made turbidity currents in Lake Superior: Sedimentology, v. 23, p. 815-831.

——— and D. J. W. Piper, 1969, Deep-sea fan-valleys, past and present: Geol. Soc. America Bull., v. 80, p. 1859-1866.

Pilkey, O. H., J. V. A. Trumbull, and D. M. Bush, 1978, Equilibrium shelf sedimentation, Rio de la Plata shelf, Puerto Rico: Jour. Sed. Petrology, v. 48, p. 389-400.

Reimnitz, E., 1971, Surf-beat origin for pulsating bottom currents in the Rio Balsas submarine canyon, Mexico: Geol. Soc. America Bull., v. 82, p. 81-90.

——— L. J. Toimil, and F. P. Shepard, 1975, Large-scale depositional and erosional feature from seismic profiles of the Rio Balsas Delta and canyons, Mexico: Geol. Soc. America Abstracts with Programs (1975), Salt Lake City, Utah, p. 1242.

Ryan, W. B. F., and B. C. Heezen, 1965, Ionian Sea submarine canyons and the 1908 Messina turbidity current: Geol. Soc. America Bull., v. 76, p. 915-932.

Schick, G. B., J. D. Isaacs, and M. D. Sessions, 1968, Autonomous instruments in oceanographic research, *in* F. Alt, ed., Marine sciences instrumentation, vol. 4: New York, Plenum Press, p. 203-230.

Scruton, P. C., 1956, Oceanography of Mississippi delta sedimentary environments: AAPG Bull., v. 40, p. 2864-2952.

Sessions, M. H., and P. M. Marshall, 1971, A precision deep-sea time release: SIO Reference Ser. 71-5, Univ. California Scripps Institution of Oceanography, La Jolla, 18 p.

Shepard, F. P., 1951, Mass movements in submarine canyon heads: Trans. Am. Geophysical Union, v. 32, p. 405-418.

——— 1955, Delta-front valleys bordering the Mississippi distributaries: Geol. Soc. America Bull., v. 66, p. 1489-1498.

——— 1964, Sea-floor valleys of Gulf of California, *in* Marine geology of the Gulf of California: AAPG Memoir 3, p. 157-192.

——— 1965, Submarine canyons explored by Cousteau's diving saucer, *in* W. F. Whittard and R. Bradshaw, eds., Submarine geology and geophysics: London, Butterworths, p. 303-311.

——— 1966, Meander in valley crossing a deep-sea fan: Science, v. 154, p. 385-386.

——— 1973, Sea floor off Magdalena delta and Santa Marta area, Colombia: Geol. Soc. America Bull., v. 84, p. 1955-1972.

——— 1975, Progress of internal waves along submarine canyons: Marine Geology, v. 19, p. 131-138.

——— 1977, Geological oceanography: evolution of coasts, continental margins, and the deep-sea floor: New York, Crane, Russak & Co., 214 p.

——— and R. F. Dill, 1966, Submarine canyons and other sea valleys: Chicago, Rand McNally & Co., 381 p.

——— ——— 1977, Currents in submarine canyon heads off north St. Croix, U.S. Virgin Islands: Marine Geology, v. 24, p. M39-M45.

——— and K. O. Emery, 1973, Congo submarine canyon and fan valley: AAPG Bull., v. 57, p. 1679-1691.

——— and N. F. Marshall, 1973, Storm-generated current in a La Jolla submarine canyon, California: Marine Geology, v. 15, p. M19-M24.

——— ———, 1978, Currents in submarine canyons and other sea valleys, chapter 1, *in* D. J. Stanley and G. Kelling, eds., Sedimentation in submarine canyons, fans, and trenches:

Stroudsburg, Pa., Dowden, Hutchinson & Ross, Inc., p. 3-14.

——— and J. D. Milliman, 1978, Sea-floor currents on the foreset slope of the Fraser River delta, British Columbia (Canada): Marine Geology, v. 28, p. 245-251.

——— R. F. Dill, and U. von Rad, 1969, Physiography and sedimentary processes of La Jolla submarine fan and fan valley, California: AAPG Bull., v. 53, p. 390-420.

——— N. F. Marshall, and P. A. McLoughlin, 1974, Currents in submarine canyons: Deep-Sea Research, v. 21, no. 9, p. 691-706.

——— ——— ———, 1975, Pulsating turbidity currents with relationship to high swell and high tides: Nature, v. 258, p. 704-706.

——— R. R. Revelle, and R. S. Dietz, 1939, Ocean-bottom currents off the California coast: Science, v. 89, p. 488-489.

——— et al, 1976, Sediment waves (giant ripples) transverse to the west coast of Mexico: Marine Geology, v. 20, p. 1-6.

——— et al, 1977, Current-meter recordings of low-speed turbidity currents: Geology, v. 5, p. 297-301.

Shields, A., 1936, Anwendung der Achnlichkeitsmechanik und der Turbulenz-forschung auf die Geschiebebewegung: Mitt. Preuss. Versuchsanst. Wasserbau und Schiffbau, 26.

Stanley, D. J., 1974, Pebbly mud transport in the head of Wilmington Canyon: Marine Geology, v. 16, p. M1-M8.

——— and G. Kelling, 1967, Sedimentation patterns in the Wilmington submarine canyon area, *in* Ocean sciences and engineering of the Atlantic Shelf: Trans. Natl. Symposium Marine Tech. Soc. (Philadelphia).

——— P. Fenner, and G. Kelling, 1972, Currents and sediment transport at the Wilmington Canyon shelfbreak, as observed by underwater television, *in* Swift, Duane, and Pilkey, eds., Shelf sediment transport: Stroudsburg, Pa., Dowden, Hutchinson & Ross Inc.

Sternberg, R. W., 1965, Observations of boundary layer flow in a tidal current: Thesis, Univ. Washington, Seattle, 71 p.

——— 1967, Measurements of sediment movement and ripple migration in a shallow marine environment: Marine Geology, v. 5, p. 195-205.

——— 1971, Measurements of incipient motion of sediment particles in the marine environment: Marine Geology, v. 10, p. 113-119.

——— and J. S. Creager, 1965, An instrument system to measure boundary-layer conditions at the sea floor: Marine Geology, v. 3, p. 475-482.

Stetson, H. C., 1936, Geology and paleontology of the Georges Bank canyons, part 1, geology: Geol. Soc. America Bull., v. 47, p. 339-366.

——— 1938, The sediments of the continental shelf off the eastern coast of the United States: Papers in Physical Oceanography and Meteorology, Cambridge, Mass., M.I.T. and W.H. O.I., v. 5, p. 5-48.

——— 1949, The sediments and stratigraphy of the East Coast continental margin; Georges Bank to Norfolk Canyon: Papers in Physical Oceanography and Meteorology, Cambridge, Mass., M.I.T. and W.H.O.I., v. 11, 60 p.

Tiffin, D. L., et al, 1971, Structure and origin of foreslope hills, Fraser delta, British Columbia: Bull. Canadian Petroleum Geology, v. 19, p. 589-600.

Wright, L. D., and J. M. Coleman, 1971, Effluent expansion and interfacial mixing in the presence of a salt wedge, Mississippi River delta: Jour. Geophys. Research, v. 76, p. 8649-8661.

Yerkes, R. F., D. S. Gorsline, and G. A. Rusnak, 1967, Origin of Redondo submarine canyon, Southern California: U.S. Geol. Survey Prof. Paper 575-C, p. C97-C105.

Explanation of Indexing

A reference is indexed according to its important, or "key" words.

Three columns are to the left of the keyword entries. The first column, a letter entry, represents the AAPG book series from which the reference originated. In this case, S stands for Studies in Geology Series. Every five years, AAPG merges all its indexes together, and the letter S will differentiate this reference from those of the AAPG Memoir Series (M) or from the AAPG Bulletin (B).

The following number is the series number. In this case, 8 represents a reference from Studies No. 8.

The last column entry is the page number in this volume where the reference will be found.

Note: This index is set up for single-line entry. Where entries exceed one line of type, the line is terminated. (This is especially evident with manuscript titles, which tend to be long and descriptive.) The reader sometimes must be able to realize keywords, although commonly taken out of context.